SLUDGE MANAGEMENT & DISPOSAL

FOR THE PRACTICING ENGINEER

SLUDGE MANAGEMENT & DISPOSAL

FOR THE PRACTICING ENGINEER

P. Aarne Vesilind
Gerald C. Hartman
Elizabeth T. Skene

LEWIS PUBLISHERS, INC.

Library of Congress Cataloging-in-Publication Data

Vesilind, Aarne P.
    Sludge management and disposal — for the practicing engineer.

    Includes bibliographies and index.
    1. Sewage sludge — United States. 2. Sewage disposal —
Florida — Case studies. I. Hartman, Gerald C. II. Skene,
Elizabeth. III. Title.
TD767.V46   1985   628.3′6   85-24189
ISBN 0-87371-060-6

COPYRIGHT © 1986 by LEWIS PUBLISHERS, INC.
ALL RIGHTS RESERVED

LEWIS PUBLISHERS, INC.
121 South Main Street, P.O. Drawer 519, Chelsea, Michigan 48118

PRINTED IN THE UNITED STATES OF AMERICA

PREFACE

The wastewater treatment profession is unique in that it has little control over the raw material it has to process. With some minor exceptions, such as industrial pretreatment requirements and sewer surcharges, the wastewater treatment plant operator literally must accept whatever comes down the pipe, and treat it so as to produce a clean water which can be discharged into the environment.

But a wastewater treatment plant produces another "effluent" — sludge. Obviously, with such a varied input of raw materials and a wide variety of available treatment technologies, the characteristics and quantities of wastewater treatment plant sludges vary markedly. Sludge management, which includes treatment of sludge for the removal or destruction of unwanted constituents, sludge dewatering, and finally ultimate disposal, must therefore be site specific. That is, every treatment plant will produce different sludges in different amounts, and the sludge handling system must be able to accommodate the specific sludges produced in that treatment plant.

For this reason (primarily) there exist many alternatives for sludge treatment. Any number of systems are available for sludge stabilization, for example, and the selection depends on site-specific criteria. Thus, since each sludge and facility has unique characteristics, there is no universally acceptable "correct" way of managing sludge. The decision as to what system to use depends on many variables, often including political considerations and preconceived and perhaps biased notions of what does and does not work.

The consulting engineer, when approached to perform a study of sludge management options, is confronted by all of the above problems. The engineer must negotiate through the maze of technological alternatives, economic calculations and local prejudices, and develop a workable and reasonable alternative for the client. The primary objective of this book is to assist engineers in this endeavor.

This book is divided into two parts. Part I is a review of the considerations that engineers must understand when starting a sludge management study. Included in this part is a section on regulatory considerations. Since this book is intended mainly for engineers working in the United States, only federal and individual state regulations are addressed. It is recognized that many other nations have developed well reasoned and locally applicable sludge disposal standards, but space precludes a discussion of these.

The second part of the book is a case study. This study was conducted for the City of St. Petersburg, Florida, by the engineering firm of Dyer, Riddle, Mills and Precourt, Inc. of Orlando, Florida. Because St. Petersburg presents many of the problems associated with sludge disposal, it is an excellent example of how the principles outlined in Part I of this book are applied in practice. This case study also includes information on available technologies and costs of alternatives, and this information can be of great assistance to other engineers embarking on similar studies for their clients.

The enthusiastic encouragement of the City of St. Petersburg is greatly appreciated. The authors wish to especially acknowledge the assistance of Mayor Corrinne Freeman; Jordan M. Rich, Administrator, Public Works; William D. Johnson, Director, Public Utilities; Robert E. Bedell, P.E., Engineering Director; A. Glenn Greer, P.E., Assistant Director, Public Utilities; J. David Shulmister, Wastewater Operations Manager; and Lee Harrison, Wastewater Chemist.

P. Aarne Vesilind
Gerald C. Hartman
Elizabeth T. Skene

This book is dedicated to

Aino Vesilind

Lynne and Christopher Hartman

George, Andrea and Cody Skene

P. Aarne Vesilind, P.E. is Professor and Chairman, Department of Civil and Environmental Engineering at Duke University, Durham, North Carolina, and has been Director of the Duke Environmental Engineering Center since 1974. Born in Estonia, he holds BS and MS degrees in civil engineering from Lehigh University, and an MS and PhD in sanitary engineering from the University of North Carolina. Dr. Vesilind was a Fellow at the Norwegian Institute for Water Research, was a visiting lecturer at Danish Technology University, and was a Fulbright-Hayes Senior Lecturer at the University of Waikato in Hamilton, New Zealand.

His numerous awards and honors include the Duke University Outstanding Professor Award for 1971–72, the 1971 Collingwood Prize from the American Society of Civil Engineers in 1971, the Earl I. Brown Outstanding Civil Engineering Teaching Award in 1982, and the Special Service Award E-38 Committee on Resource Recovery from the American Society for Testing and Materials in 1984. He is active in a number of professional societies.

Professor Vesilind has done design and consulting work for industrial and consulting firms, and for government agencies, and has an impressive record of sponsored research. He is a Registered Professional Engineer in North Carolina.

Author or co-author of 12 books, his eleventh and twelfth volumes are *Sludge Management and Disposal* (with Hartman and Skene) and *Environmental Ethics for Engineers* (with Alastair Gunn). He is the author of more than 100 research papers and articles, and serves on the Editorial Board of Lewis Publishers, Inc.

Gerald C. Hartman, P.E. is vice president and partner of Dyer, Riddle, Mills & Precourt, Inc. Mr. Hartman was a pupil of Dr. Vesilind both in undergraduate and graduate school. Mr. Hartman has served as a consultant to numerous municipalities, industries and agricultural concerns in the Southeast United States on sludge management and disposal. He has served as a technical expert witness for environmental litigation, and has written several technical articles in the sanitary field. Moreover, he is the design engineer of record for a number of water and wastewater facilities in the Southeast.

Elizabeth T. Skene is currently the Environmental Compliance Manager with the City of Orlando, Florida. She holds a BS degree from the University of South Florida and a MS degree from the University of Central Florida's Civil Engineering and Environmental Sciences Department. Ms. Skene is a member of Tau Beta Phi, American Society of Civil Engineers, Water Pollution Control Federation, and numerous other organizations. She was chosen as Outstanding Young Woman of 1984. With the City of Orlando, she is investigating the effect of industrial pretreatment on sludge quality.

CONTENTS

Preface.. v
List of Tables xv
List of Figures.. xix
List of Technology Profiles and Installation Surveys........... xxiii

PART ONE
SLUDGE DISPOSAL ALTERNATIVES, AVAILABLE TECHNOLOGY AND REGULATIONS

1. **Estimating Sludge Production and Sludge Characteristics**..... **3**
   Introduction... 3
   Sludge Characterization................................ 4
   Estimating Sludge Quantities 9

2. **Sludge Stabilization Options**........................... **15**
   Introduction... 15
   Defining Sludge Stability.............................. 15
   Measuring Sludge Stability 17
   Stabilizing Options 21
   Description of Sludge Stabilization Options............ 22

3. **Sludge Dewatering Technology**.......................... **39**
   Introduction... 39
   Sludge Dewatering Options.............................. 39

4. **Sludge Disposal**...................................... **63**
   Introduction... 63
   Landfilling 63
   Landspreading on Agricultural Land..................... 64
   Landspreading on Reclaimed Land 67
   Land Farming 67
   Ocean Disposal 70
   Deep Well Injection 71
   Contract Disposal 72

5. **Evaluation of Alternatives**................................ **73**
   Introduction.. 73
   Transformation Curves 73

6. **Regulatory Constraints**................................ **79**
   Introduction.. 79
   Applicable Federal Laws 79
   Applicable Federal Regulations 81

7. **Conclusions** **85**
   Introduction.. 85
   Three Laws of Sludge Management.................... 85
   Holding Sludge 86
   Mixing Sludge 86
   Recirculating Sludge............................... 87
   Conclusions 87

PART TWO
SLUDGE DISPOSAL FOR ST. PETERSBURG: A CASE STUDY

1. **Introduction**.. **91**
   Purpose and Scope.................................. 92
   Organization 92
   Methodology for Technology Assessment.............. 95

2. **Background** **97**
   General Background................................. 98
   Previous Studies 98
   Summary of Applicable Sludge Rules and Regulations 101
   Existing Facilities 107

3. **Sludge Quantities and Characteristics** **117**
   Existing Sludge Quantities.......................... 118
   Existing Sludge Characteristics...................... 121
   Projections.. 124

4. **Available Technology** **131**
   Introduction....................................... 132
   Sludge Drying 132
   Combustion 137
   Composting and Earthworm Conversion 142
   Ultimate Sludge Disposal........................... 147

5. **Technology Assessment** **169**
   General 170
   Summary of Sludge Disposal Alternatives 171
   Stabilization/Disinfection Considerations.................. 179

6. **Alternative Approaches** **185**
   Introduction... 186
   City of St. Petersburg Objectives.......................... 186
   Project Constraints... 187
   Contract Hauling and Disposal 188
   City-Owner/Operator.. 192
   City/County Regional Incineration Project 194
   Full Service Operator 197

7. **Development of Alternatives**............................. **201**
   Introduction... 202
   Description of Alternatives 206
   Costing Factors for Capital and O&M Cost Estimates....... 211
   Cost Estimates of Alternatives 213
   Survey of Technical Processes Used in the Ten Alternatives
      for Sludge Management 245

8. **Comparative Analysis of Alternatives Utilizing**
   Transformation Curves **253**
   General 254
   Transformation Curve Analysis 254
   Evaluation Factors/Considerations 254
   Comparative Evaluations.................................... 263
   Summary of Comparatively Favorable Alternatives 263

9. **User Fees and Payout Period Analyses of the Favorable**
   Alternatives **275**
   General 276
   User Fee Analysis 276
   Payout Period Analyses..................................... 289
   Recommendations... 292

10. **Conclusions and Recommendations**....................... **299**
   Summary of the Report 300
   Conclusions 300
   Recommendations.. 302
   Implementation Schedule.................................... 303

11. **Project Update** **305**
    General 306
    Technology Update of Recommended Alternatives.......... 306

**Appendix: Florida Department of Environmental Regulation
Chapter 17-7 Regulations** **309**

LIST OF TABLES

PART ONE

1.1 Ranges of Sludge Production in Europe 10
1.2 Types of Sludge Production in Europe 10
1.3 Split Factors for Estimating Sludge Production 13
2.1 Effectiveness of Sludge Stabilization Processes 21
2.2 Typical Digester Design Criteria 24
2.3 Bacterial Survival in Digestion 26
2.4 Aerobic Digestion Design Parameters 28
2.5 Representative Heating Values of Some Sludges 33
3.1 Techniques for Achieving Solid/Liquid Separation for
    Wastewater Sludges 40
3.2 Design and Operational Variables for Solid Bowl Centrifuges   54
4.1 Nitrogen and Phosphorus in Sludge and Comparable
    Organic Wastes 65
4.2 Cadmium in Sludge 66
4.3 Surface and Subsurface Application Methods and Equipment
    for Liquid Sludges 68
4.4 Methods and Equipment for Application of Dewatered
    Sludges 71
5.1 Summary of Transformation Curve Analysis Shown in
    Figure 5.2 75

PART TWO

2.1 Classification Parameters for Sludges — State of Florida
    Department of Environmental Regulation 104
2.2 Permitting Requirements Under FDER 17-7 Part IV for
    Land Application FAC 17-7.54(4) and 17-4.64 105
2.3 State of Florida Ambient Air Quality Standards 106
2.4 Summary of City of St. Petersburg Treatment Facilities 109

2.5 Existing Sludge Facilities at the City of St. Petersburg
      Wastewater Treatment Plants........................   110
3.1 Wastewater Treatment Plant Sludge Production Statistics....   119
3.2 Sludge Production Before and After Belt Filter Press
      Dewatering   120
3.3 Sludge Toxicity Analysis   122
3.4 Typical Chemical Composition of Digested Sludge..........   123
3.5 Comparison of Existing and Projected Wastewater Flows....   125
3.6 Projected Daily Sludge Production for the City of St.
      Petersburg   127
4.1 Combustion processes   138
4.2 Other High Temperature Processes   139
4.3 Some In-Vessel Composting Systems Marketed in the United
      States   144
4.4 Sludge Solids Content and Handling Characteristics   152
4.5 Transport Modes for Sludges   153
5.1 Preliminary Screening Matrix—*Sludge Drying*.............   172
5.2 Preliminary Screening Matrix—*Combustion Methods*   173
5.3 Preliminary Screening Matrix—*Composting and Others*   174
5.4 Preliminary Screening Matrix—*Ultimate Sludge Disposal*....   175
5.5 Preliminary Screening Matrix—*Modes of Transportation for
      Contract Hauling*   176
5.6 Relative Effects of Various Treatment Processes on
      Destruction of Pathogens and Stabilization of Sewage
      Sludges..   181
5.7 Stabilization/Disinfection Potential by the Various
      Technologies   183
7.1 Selected Alternatives for Further Comparative Analysis   204
7.2 Alternative A Estimated Project Cost Rotary Drying/Full
      Service Operator....................................   217
7.3 Alternative A Estimated Operation & Maintenance Cost
      Rotary Drying/Full Service Operator   218
7.4 Alternative B Estimated Project Cost Multiple-Effect
      Drying/Full Service Program.........................   222
7.5 Alternative B Estimated Operation & Maintenance Cost
      Multiple-Effect Drying/Full Service Operator...........   223
7.6 Alternative C Estimated Project Cost Solar Powered
      Drying/Full Service Operator.........................   225
7.7 Alternative C Estimated Operation & Maintenance Cost
      Solar Powered Drying/Full Service Operator   227
7.8 Alternative D Estimated Project Cost Fluidized Bed
      Incinerator/Full Service Operator.....................   229

7.9 Alternative D Estimated Operation & Maintenance Cost FY
    1987 Fluidized Bed Incinerator/Full Service Operator..... 231

7.10 Alternative E Estimated Project Cost In-Vessel
    Composting/Full Service Operator 235

7.11 Alternative E Estimated Operation & Maintenance Cost FY
    1987 In-Vessel Composting/Full Service Operator 236

7.12 Alternative F Estimated Project Cost Lime
    Encapsulation/Full Service Operator 239

7.13 Alternative F Estimated Operation & Maintenance Cost
    Lime Encapsulation/Full Service Operator 240

7.14 Alternative G Summary of Costs City/County Incineration
    Project 242

7.15 Alternative H Estimated Operation & Maintenance Cost
    Lime Encapsulation/City Owned and Operated 244

8.1 Selection Criteria Categories 255

8.2 Operation and Maintenance Criteria Comparative Evaluation 257

8.3 Welfare of the Community Comparative Evaluation 258

8.4 Other Engineering Considerations Comparative Evaluation .. 259

8.5 Summary Comparative Evaluation and Costs for the
    Selected Alternatives 260

8.6 Results of the Transformation Curves.................... 271

8.7 The Most Favorable Sludge Disposal Alternatives 272

9.1 Alternative B—Projected Disposal Costs Multiple-Effect
    Drying/Full Service Operator......................... 278

9.2 Alternative E—Projected Disposal Costs In-Vessel
    Composting/Full Service Operator 280

9.3 Alternative G—Projected Disposal Costs City/County
    Incineration Project................................. 282

9.4 Alternative H—Projected Disposal Costs Lime
    Encapsulation 286

9.5 Alternative I—Projected Disposal Costs In-Vessel
    Composting with City Ownership 288

9.6 Alternative J—Contract Hauling/Truck Projected Disposal
    Costs.. 290

9.7 Summary of Favorable Alternatives 291

9.8 Annual Costs used for Payout Period Analysis............. 297

10.1 Implementation Schedule City of St. Petersburg Sludge
    Disposal Program................................... 304

LIST OF FIGURES

PART ONE

1.1 Description of sludge by particle size.................... 6

1.2 Description of water in sludge, as extracted by centrifugal acceleration in a test tube centrifuge.................. 6

1.3 A rheogram for municipal sludge...................... 8

2.1 Sludge stabilization processes are defined as those processes which, recognizing a potential adverse characteristic of the wastewater sludge, change that characteristic so as to make disposal of the sludge acceptable................ 16

2.2 Relationship between CO_2 on the digester gas and bicarbonate alkalinity............................... 23

2.3 Typical two-stage anaerobic digesters.................... 24

2.4 Aerobic digestion...................................... 27

2.5 Schematic representing the basic processes in sludge incineration. 31

2.6 Multiple-hearth incinerator. 31

2.7 Fluid bed incinerator................................... 32

2.8 Process zones in a multiple-hearth incinerator............ 32

2.9 Static aerated pile composting.......................... 34

2.10 Typical enclosed-vessel composting plant. 35

2.11 Destruction of pathogens in a static pile composting operation. 36

3.1 Three basic principles used in solid/liquid separation....... 40

3.2 Gravity thickener cross section. 41

3.3 Calculation of limiting flux. 43

3.4 Effect of initial depth and cylinder diameter on settling velocity for a typical activated sludge at 2000 mg/l. 44

3.5 Dissolved air flotation thickener....................... 46

3.6 Flotation apparatus..................................... 47

3.7 The air-to-solids ratio affects both the thickened solids and the recovery of solids in a dissolved air flotation system. . 47

3.8 Solid bowl centrifuge. 49
3.9 Solids recovery of a dilute and concentrated calcium
     carbonate slurry.................................. 51
3.10 A strobe light placed above a spinning test tube can be used
     to observe the settling of a sludge under gravitational
     force. 52
3.11 Penetrometer used for estimating the firmness of a sludge. . 53
3.12 Centrifuge performance tradeoff between cake solids
     concentration and solids recovery..................... 55
3.13 Rotary drum vacuum filter............................ 56
3.14 The belt filter employs gravity drainage and positive
     pressure... 56
3.15 The filter press has plates and frames which are covered by
     a filter cloth and through which the liquid flows as the
     sludge is pumped in under pressure................... 57
3.16 Filter leaf apparatus used for sizing vacuum filters........ 58
3.17 Buchner funnel apparatus for determining the specific
     resistance to filtration. 59
3.18 Typical results of a Buchner funnel test for measuring
     specific resistance to filtration. 59
3.19 Capillary suction time apparatus....................... 60
3.20 Cross section of a standard sand drying bed. 61
4.1 Cadmium balance in solids. 66
5.1 Typical transformation curve........................... 74
5.2 Transformation curves with multiple objectives........... 76

PART TWO

1.1 Graphic representation of sludge study. 93
1.2 Study area....................................... 94
2.1 Existing St. Petersburg facilities. 108
2.2 Existing method of sludge disposal..................... 114
4.1 Basic sludge disposal alternatives. 133
7.1 Summary of sludge disposal options. 203
7.2 Proposed site plan for rotary drying at the N.E. WWTP..... 214
7.3 Proposed site plan for rotary drying at the S.W. WWTP..... 215
7.4 Proposed site plan for multi-effect drying at the N.E.
    WWTP. 220
7.5 Proposed site plan for multi-effect drying at the S.W.
    WWTP. 221
7.6 Proposed site plan for city only incinerator. 228

7.7 Proposed site plan for in-vessel composting at the N.E. WWTP.   233

7.8 Proposed site plan for in-vessel composting at the S.W. WWTP.   234

7.9 Typical layout for lime encapsulation installation.   238

8.1 Operation and maintenance versus capital cost.   264

8.2 Welfare of the community evaluation versus capital cost.   265

8.3 Other engineering considerations versus capital cost.   266

8.4 Operation and maintenance considerations versus operation and maintenance cost.   267

8.5 Welfare of the community versus operation and maintenance cost.   268

8.6 Other engineering considerations versus operation and maintenance cost.   269

8.7 Operation and maintenance cost versus capital cost.   270

9.1 Projected total annual cost.   334

9.2 Payout period for regional incinerator versus contract hauling.   335

9.3 Payout period for in-vessel composting FSO versus contract hauling.   336

9.4 Payout period for lime encapsulation versus contract hauling.   337

TECHNOLOGY PROFILES AND INSTALLATION SURVEYS

4.1 Flash Drying Technology Profile 155
4.2 Rotary Drying Technology Profile...................... 156
4.3 Multiple-Effect Drying Technology Profile 157
4.4 Solar Powered Drying Technology Profile 158
4.5 Multiple Hearth Incineration Technology Profile 159
4.6 Fluidized Bed Incineration Technology Profile 160
4.7 Electric (Infrared) Furnace Incineration Technology Profile. 161
4.8 Co-Combustion Technology Profile 162
4.9 In-Vessel Composting Technology Profile 163
4.10 Windrow Composting Technology Profile 164
4.11 Aerated Pile Composting Technology Profile 165
4.12 Earthworm Conversion Technology Profile 166
4.13 Lime EncapsulationTechnology Profile.................. 167

7.1 Sludge Treatment Installation Survey 246
7.2 Sludge Treatment Installation Survey 247
7.3 Sludge Treatment Installation Survey 248
7.4 Sludge Treatment Installation Survey 249
7.5 Sludge Treatment Installation Survey 250
7.6 Sludge Treatment Installation Survey 251

PART I

SLUDGE DISPOSAL ALTERNATIVES, AVAILABLE TECHNOLOGY AND REGULATIONS

CHAPTER 1

ESTIMATING SLUDGE PRODUCTION AND SLUDGE CHARACTERISTICS

INTRODUCTION

The treatment of municipal wastewater results in the formation of slurries high in suspended solids. These slurries, commonly referred to as sludges, are produced either by the concentration of solids originally in the wastewater (such as raw primary sludge) or the formation of new suspended solids as the result of removing dissolved solids from the wastewater (such as waste activated sludge). At times these sludges can be disposed of into the environment without further treatment. For example, the spraying of waste activated sludge or sludges from fruit canning plants into forests is a well established practice and has a net beneficial impact on the environment since these sludges are readily assimilated into the forest ecosystem and cause no detrimental perturbations.

Too often, however, it is not feasible, either environmentally or economically, to dispose of such sludges directly into the environment or to use them for other beneficial purposes. Raw primary sludge, for example, is odiferous, full of large solids and pathogenic organisms, and can seldom be discharged onto land without some prior treatment. For many municipal sludges, treatment may involve the removal of some of the liquid to make it more economically transportable, destruction of pathogenic organisms which might cause public health problems, reduction of the nuisance conditions, especially problems with odor, and removal of other unwanted constituents.

Before such treatment can be considered, and prior to any considerations of ultimate disposal alternatives, sludges must be characterized according to the properties of importance, and the quantities of sludges to be handled must be approximated or measured. The objective of this section is to develop a rational basis for sludge characterization, and to provide methods for the estimation of sludge production.

SLUDGE CHARACTERIZATION

Municipal wastewater treatment plant sludges are most readily characterized by their source, such as primary sludge, or waste activated sludge. This stereotyping is useful because often sludges from similar sources exhibit similar characteristics, and these may be fairly predictable. For example, waste activated sludges are known to be full of microbial life, are notoriously difficult to dewater, and seldom thicken to solids concentrations greater than 2% solids. Having said that, it is necessary to point out that one could readily encounter a waste activated sludge that is microbiologically inert and dewaters like a sieve. The usefulness of sludge stereotyping on the basis of source is only marginally beneficial, and exceptions should be expected. Below are some other useful methods of characterizing sludges which rely on analytical measurements and thus provide numerical information which can be of value in the design of sludge management alternatives.

Solids Concentration

A sludge is defined as a slurry containing a solid phase suspended in a liquid. A primary characteristic of sludges is the ratio of the solids to the liquid. In modern environmental engineering practice, solids concentrations are expressed as

$$C_1 = \frac{\text{mg dry solids}}{\text{liters of sludge}} = \text{mg/L}$$

It should be noted that this is not mg of solids per liter of *water* (or whatever liquid the solids are suspended in). When the concentrations of solids are sufficiently high that the tests for suspended solids as specified by *Standard Methods* (16th ed.) or the American Society for Testing and Materials (ASTM) no longer are possible, solids concentrations are expressed as

$$C_2 = \frac{\text{g dry solids}}{\text{g sludge}} = \text{g/g}$$

This number multiplied by 100 yields the oft used "percent solids." Note that this equation does not calculate g dry solids per g of water.

The first equation is a mass/volume relationship, while the second is a mass/mass relationship. If it is assumed that the density of the solids is

1.00, the same as water, the two expressions can be related as

$$C_1 = mg/L = 10,000 \times \text{percent solids} = C_2$$

Repeating: this expression holds *only if the density of the sludge is assumed to be 1.00, the same as water*. When handling slurries composed of industrial suspended solids such as flue gas desulfurization sludges, for example, this relationship does not hold since the density of the dry solids is significantly higher than that of water.

In this book, the assumption is made that the above expression holds, and conversions are made without further disclaimers. That is,

$$C = C_1 = C_2$$

Particle Size

It has been known for some time that the size of particles will directly affect how well the sludge will dewater. The pioneering work of Karr[1] showed that particles ranging from 100 to 1 μm seem to have a direct effect on dewaterability. Thus, if a sludge is composed of a large number of particles in this range, its dewatering will be difficult, and thus it would be advantageous to change its particle size distribution. Kavanaugh et al.[2] have used a power law,

$$\frac{dN}{dL} = AL^{-\beta}$$

which plots as a straight line with the log of the particle size vs log $\triangle N / \triangle L$, where N = particle number density and L = particle size (Figure 1.1). The factors A and β are characteristic of the particle size distribution. If $\beta = 1$, the slurry is comprised of particles of one size (uniform), and as β increases, the nonuniformity increases.

Distribution of Water

Another means of characterizing basic sludge properties is by means of how the water is attached to the particles. A number of researchers have developed schemes for classifying such water, the idea dating back to the work by Heukelekian at Rutgers in the 1930s.[3] One scheme that is useful, shown in Figure 1.2, divides water in sludge into four categories:[4]

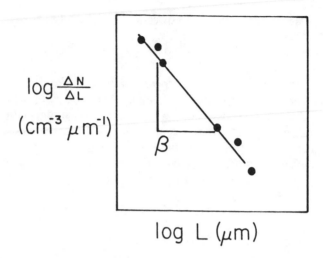

Figure 1.1. Description of sludge by particle size. L is particle diameter, and N is the number of particles of that diameter. A and beta are constants. dN/dL = AL–β; log ($\triangle$N/$\triangle$L′) = –β logL + logA.

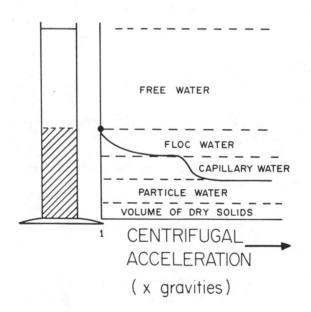

Figure 1.2. Description of water in sludge, as extracted by centrifugal acceleration in a test tube centrifuge.

1. *Free water* — the water that escapes when the sludge settles and the particles, under their own weight, force the water out. This water is not associated in any way with the particles themselves. In wastewater treatment, free water is removed by gravitational thickening.
2. *Floc water* — the water trapped in the flocs, removed when the flocs are squeezed so that they expel the water trapped in the lattice structure. Floc water is removed by mechanical dewatering.
3. *Capillary water* — the water which is attached to the particle lattice by capillary forces. This water will not be removed by mechanical means unless extreme pressures are used.
4. *Particle water* — the water chemically attached to the particle, removed only by chemical or thermal means, and only if the particle is altered (e.g., a microorganism is killed).

These are of course arbitrary distinctions, and better categorization is urgently needed. It is known, for example, that the compaction of slurries under centrifugal force occurs in stages with respect to centrifugal acceleration. There are obviously lattice structures which provide support and which are collapsed as the centrifugal force is increased,[5] and the rate at which this progressive collapsing takes place may be a useful parameter to describe sludge compaction.

Flow Properties

A potentially powerful technique in characterizing sludges is to classify them according to their physical flow properties. Colin[6] has suggested that sludge be characterized into four basic states:

1. *Liquid* — where the sludge flows under the influence of gravity.
2. *Plastic* — where the sludge is too concentrated to flow freely but permanent deformation takes place if sufficient pressure is applied. These sludges can be pumped.
3. *Solid with shrinkage* — where the sludge is too concentrated to be pumped, and its volume decreases as it dries.
4. *Solid without shrinking* — where the sludge is no longer saturated with water and dries without further shrinkage.

In dewatering, we are really only concerned with the first physical state, where the sludge acts as a liquid.

Rheological Properties

As early as 1936, Hatfield[7] recognized that wastewater sludges have curious flow properties which can be mathematically described using a power law. Dick and Ewing[8] extended this idea and suggested that sludge can be considered to be a non-Newtonian pseudo-plastic fluid, adequately described by the equation

$$\tau = \tau_o + k\gamma$$

where

$$
\begin{aligned}
\tau &= \text{shear stress (dryness/cm}^2) \\
\tau_o &= \text{yield stress (dryness/cm}^2) \\
k &= \text{apparent viscosity} \\
\gamma &= \text{shear rate (sec}^{-1})
\end{aligned}
$$

as shown in Figure 1.3. More recently, Campbell and Crescuolo[9] in Canada have found that a more accurate representation of a sludge rheogram can be obtained by using the equation

$$\log \gamma = a + n \log \mu$$

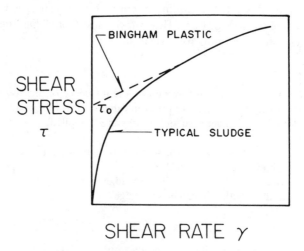

Figure 1.3. A rheogram for municipal sludge.

where

$$\gamma = \text{shear rate}$$
$$\mu = \text{viscosity at the specific shear rate}$$
$$\text{a and n} = \text{constants which depend on solids}$$
$$\text{concentration and percent volatile solids}$$

Dewatering Properties

Of interest to many engineers and sludge managers is the ability of a sludge to release water. These characteristics are discussed in detail under the sludge dewatering heading.

Stability

Similarly, sludge stability is of primary concern in sludge disposal. These properties are discussed in the stability chapter.

ESTIMATING SLUDGE QUANTITIES

Estimating expected sludge quantities is not a problem for engineers and operators of existing wastewater treatment plants. For plant expansions and for brand new plants, it is of great importance to correctly predict the quantities and characteristics of sludge produced.

The *overdesign* of sludge handling equipment can often lead to as serious operational problems as underdesign. Interestingly, because of preoccupation with the liquid stream, great difficulties often arise with sludge quantity predictions. One reason is that the quantities vary greatly from place to place. In a recent paper presented at a European sludge conference, for example, the sludge production figures for the different west European countries were enumerated (Tables 1.1 and 1.2) but the *reasons* for such wide variations were not advanced.[6] The wide range represented by these numbers makes the use of such data highly risky. Of even greater risk is the use of textbook factors which may be many years out of date.

Numbers such as those in Table 1.1 have large variation because of a number of reasons, including the following:

- Some sludge production figures are quoted on the basis of population served, while others are on the basis of population equivalent. Usually, population equivalent is expressed in terms of biological oxygen demand (BOD), but for many types of industry a high BOD does not necessarily mean high solids production.

- Some figures are reported in terms of sludge produced, while others are in terms of sludge disposed of.
- Sources of sludge such as septic tank sludge may or may not be included.
- Design standards may differ from country to country, such as mean cell residence time (sludge age), which will influence secondary sludge production.
- Treatment systems are not always well defined. An oxidation ditch, for example, may be operated with or without an anoxic region, and this will influence the amount of solids produced.
- Sewerage systems differ, with a high fraction of combined sewers (or leaky separate sewers) producing more solids per capita.
- The measurement of sludge production in a treatment plant is not a simple matter and the figures reported may be the best guesses by the operators.

Table 1.1. Ranges of Sludge Production in Europe

Countries	Sludge Produced (dry g/person/day)
The Netherlands, Ireland	30–40
Belgium, Denmark, France, Italy, Luxembourg	70–85
Finland, Norway, Sweden, Switzerland	97–124

Source: Reference 10.

Table 1.2. Types of Sludge Production in Europe

Type of Sludge	Range of Reported Sludge Production (dry g/person/day)
Raw primary sludge	36–77
Waste activated sludge	5–40
Chemical sludge	12–680

Source: Reference 10.

For a given treatment plant, it should be possible to break down the process into manageable components and develop the overall sludge production figures from a synthesis of their unit operations.

To illustrate this method, consider a wastewater treatment plant with primary clarification and activated sludge. First, the solids in the influent, the raw wastewater, must be determined. It would be helpful of course to have a very detailed analysis of the solids with respect to particle size, chemical composition, physical shape, etc., but these data are seldom available, nor would this be very useful information if it *were* available. It is reasonable to define two types of solids by the classical method, fixed and volatile, assuming of course that the fixed solids represent the inert fraction while the volatile solids represent biomass. Using this expedient, the influent is assumed to contain

$$X_o \text{ kg/day of fixed solids}$$

$$Z_o \text{ kg/day of volatile solids}$$

As this wastewater moves through the pretreatment step, some solids are removed as grit and screenings. The fraction of solids thus removed is defined as

$$p_x = \text{fraction of fixed solids removed in pretreatment}$$

$$p_z = \text{fraction of volatile solids removed in pretreatment.}$$

Obviously, the fixed solids not removed is thus $(1 - p_x) X_o$ and the volatile solids not removed is $(1 - p_z) Z_o$, and these enter the primary solids concentrator (clarifier). In order to estimate the amount of grit and screenings which must be disposed of, we need to know the values of p_x and p_z. The rest of the treatment plant is annotated in a similar manner. In the primary clarifier, the fraction of fixed solids removed is denoted as c_x, so that the total raw primary sludge produced is $(1 - p_x) (c_x) X_o$, whereas the amount of fixed solids escaping as the primary effluent is $(1 - p_x) (1 - c_x) X_o$.

In the biological treatment step, we often actually *produce* volatile solids, so that a yield coefficient is used, defined as

$$Y = \text{kg volatile solids produced per kg soluble BOD reduced}$$

It is important that yield be defined in terms of soluble BOD and *volatile* solids, for presumably we cannot produce inert solids. It is the utilization of the soluble high-energy organics that is responsible for the production of *most* waste biological sludge, *but not all*. Waste biological sludge also contains inert solids which escaped the primary clarifier $[(1 - p_x) (1 - c_x) X_o]$ and were removed in the secondary clarifier.

Thus far the analysis is fairly straightforward. The first real problem is encountered in the analysis of solids production/destruction/recycling in the anaerobic digester. Assuming that the anaerobic system has a net reduction in volatile solids and does not affect the fixed solids, the fixed solids discharged in the anaerobically digested sludge is

$$(1 - p_x) (c_x) (d_x) X_o$$

where d_x = fraction of fixed solids captured in the digester. But this means that $(1 - d_x)$ fraction escapes as the supernatant. This solids load, or

$$(1 - p_x) (c_x) (1 - d_x) X_o$$

is now imposed on the primary clarifier. But earlier it was assumed that the fixed solids loading was $(1 - p_x) X_o$. Obviously, this must be corrected, so that the *actual* loading is

$$(1 - p_x) X_o + (1 + p_x) (c_x) (1 - d_x) X_o$$

which of course makes the *actual* quantity of inert solids making up the raw primary sludge

$$c_x [(1-p_x) X_o + (1 - p_x) (c_x) (1 - d_x) X_o]$$

and the *actual* solids in the supernatant is thus

$$(1 - d_x) c_x [(1 - p_x) X_o + (1 - p_x) (c_x) (1 - d_x) X_o]$$

and so on. Obviously, this can go on forever. It is possible to escape from this infinite loop by deciding if the change from one loop to the next is sufficiently small (e.g., 1%) that it is reasonable to stop the calculations.

Recognize now that the centrifuge also has a feedback loop — the solids in the centrate. The two different solids' return will of course complicate matters considerably.

It should be obvious that this technique of estimating solids production in wastewater treatment requires the use of a computer. The program to perform the calculations is fairly straightforward, and the real trick is to give the split factors (p_x, d_x, c_x, etc.) proper values. Table 1.3 lists some of these, but caution is urged in the blind application of these values.

Jones and Vesilind[11] have shown that the iterative process can be avoided if the unit operations are defined as solids recycling processes. These calculations are beyond the scope of this discussion, however.

One final comment about analyzing solids movement in a wastewater

Table 1.3. Split Factors for Estimating Sludge Production

p_x	= fraction of fixed solids removed in pretreatment to screenings and grit	= 0.04
p_z	= fraction of volatile solids removed in pretreatment to screenings and grit	= 0.10
c_x	= fraction of fixed solids removed in primary clarifier, to raw primary sludge	= 0.67
c_z	= fraction of volatile solids removed in primary clarifier, to raw primary sludge	= 0.42
f_x	= fraction of fixed solids removed in final clarifier to waste activated sludge	= 0.90
f_z	= fraction of volatile solids (including biomass produced in conversion of BOD to solids) removed in final clarifier	= 0.90
d_x	= fraction of fixed solids removed in anaerobic digester as mixed digested sludge	= 0.80
d_z	= fraction of volatile solids removed in anaerobic digester as mixed digested solids	= 0.40
v_z	= fraction of volatile solids destroyed in anaerobic digestion	= 0.40
n_x	= fraction of fixed solids removed in centrifugal dewatering as dewatered sludge	= 0.90
n_z	= fraction of volatile solids removed in centrifugal dewatering as dewatered sludge	= 0.80

treatment plant is necessary: It has been assumed that the solids are adequately described by dry mass. This is of course not true, since the characteristics of these solids will influence how the process will behave. For example, suppose a great many small solids particles are recycled from the anaerobic digester as supernatant. Back in the primary clarifier they can flocculate with the incoming raw wastewater solids, be recaptured, and end up back in the anaerobic digester. If this continues, the system will be clogged up with these solids and treatment will suffer. The only solution in such a situation is not to recycle any supernatant, but to dewater the entire flow from anaerobic digestion until the system is cleaned out.

REFERENCES

1. Karr, P. "Factors Influencing the Dewatering Characteristics of Sludge," PhD Dissertation, Clemson University, 1976.
2. Kavanaugh, M. C., C. H. Tate, A. R. Trussell, R. R. Trussell, and G. Treweek. "Use of Particle Size Distribution Measurements for Selection and Control of Solid/Liquid Separation Processes," *Particulates in Water*, M. C. Kavanaugh and J. O. Leckie, Eds., Advances in Chemistry Series 189 (Washington, DC: American Chemical Society 1980).
3. Heukelekian, H., and E. Weisberg. "Bound Water and Activated Sludge Bulking," *Sew. Ind. Wastes*, 28(4) (1956).
4. Vesilind, P. A. *Treatment and Disposal of Wastewater Sludges*, (Ann Arbor, MI: Ann Arbor Science Publishers, Inc., 1978).
5. Buscall, R. "The Elastic properties of Structured Dispersions: A Simple Centrifuge Method of Examination," *Colloids and Surfaces*, 5(4):269–283 (1982).
6. Colin, F. "Characterization of the Physical State of Sludges," paper presented at the COST 68 Symposium, Brighton, England, 1983.
7. Hatfield, W. D. "The Viscosity of Pseudo-Plastic Properties of Sewage Sludge," *Sew. Works*, 10(1):3–25 (1938).
8. Dick, R. I., and B. B. Ewing. "The Rheology of Activated Sludge," *J. Water Poll. Control Fed.* 39(4):543–560 (1967).
9. Campbell, H. W., and P. J. Crescuolo. "The Use of Rheology of Activated Sludge," *J. Water Poll. Control Fed.* 39(4):543–560 (1967).
10. Poulanne, J. "Sludge Production Rates," paper presented at the COST 68 Symposium, Brighton, England, 1983.
11. Jones, G. N., and P. A. Vesilind, "Sludge Production in a Wastewater Treatment Plant," *Journal Envir. Eng. Dir.* ASCE (in press).

CHAPTER 2

SLUDGE STABILIZATION OPTIONS

INTRODUCTION

The treatment of domestic and industrial wastewaters results in the production of sludges which have uncommonly undesirable characteristics, and thus these sludges are to be further processed so as to make their disposal or use acceptable. Such treatment is commonly called "stabilization."

Although sludge stabilization has been practiced for over 100 years, and although it represents a major cost center in wastewater treatment, and although every treatment plant in the world incorporates some type of stabilization into the process, it is still difficult to clearly define what is meant by stabilization; nor is there agreement on a definition of stabilized sludge.

DEFINING SLUDGE STABILITY

The definitions of sludge stability in common use today fall into two categories:

1. Those that define how much improvement in any of one or more sludge characteristics is achieved.
2. Those that specify a treatment which will commonly attain some significant improvement in one of these characteristics.

An example of the first type of definition is the emerging sludge disposal standard written by the United States Environmental Protection Agency,[1] in which sludge is to be stabilized in order to "significantly reduce pathogens" or to reduce the volatile solids by a certain percentage. The direction and objective of these standards is to specify the effect of the process on specified sludge parameters. However, in the case of pathogens, it is important to note that some methods of pathogen reduction do not improve undesirable sludge characteristics such as odor, toxins and poor dewaterability. If the concentration of pathogens is used as the only definition of stability it is possible to produce a sludge which, although free of pathogens, cannot be disposed of without significant damage to the environment or to public health. Similarly, a reduction of volatile solids does not directly measure odor reduction, dewatering, or other important parameters.

An example of the second definition is the sludge disposal guidelines presently being written by the Concerted Action on Sludge Disposal (COST 68), organized by the European Economic Community, which states that a stable sludge is one which has received "biological or chemical treatment or long-term storage."[2] It should be obvious that unless the degree of such treatment is specified, no guarantees as to the ultimate effect of sludge on the environment or on human health can be made.

In short, it does not seem rational to try to define sludge stability by either a single or selected group of parameters, or by the form of treatment offered.

A suggested alternative is to picture the sludge management process in a holistic manner, as shown in Figure 2.1. The raw sludge is in one end of this continuum, with the environment in the other. The process of stabilization is the link connecting the raw sludge with its eventual use of disposal in the environment, and the processes selected must be those that produce a product from a given raw material which will not cause unacceptable environmental damage when disposed of by a given means. Stated positively:

> Sludge stabilization is defined as a process or series of processes which produces a sludge with characteristics such that its ultimate use will be acceptable in terms of environmental impact and public health.

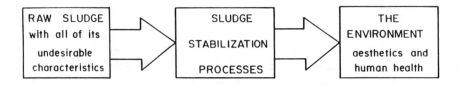

Figure 2.1. Sludge stabilization processes are defined as those processes which, recognizing a potential adverse characteristic of the wastewater sludge, change that characteristic so as to make the disposal of the sludge acceptable.

And as a corollary;

> A stable sludge is one which can be used or disposed of without undue damage
> to the environment or to public health.

The key word of course is "acceptable," which opens the way for risk
evaluation, cost trade-offs and professional judgment. These are all difficult to manage from the regulatory standpoint, and thus governmental
agencies tend to shy away from definitions based on such criteria. It nevertheless seems to be a rational definition of sludge stability, and it should be
used even if it presents regulatory difficulties. The alternative is to adopt
definitions of stability which, even though simple to enforce, may lead to
unnecessary expense in treatment, or worse, destruction of environmental
quality, or human disease and suffering.

If the definition of stabilization as stated above is adopted, what are the
sludge characteristics which might make the disposal or use of this material
unacceptable? Other than the sheer mass of the sludge, it could be argued
that there are four basic problems with waste sludges:

1. They may be odiferous.
2. They may contain pathogenic organisms.
3. They may contain chemical toxins.
4. They may be difficult and expensive to dewater.

It can then be argued that sludge stabilization processes are intended to
correct one or more of these undesirable sludge characteristics.

As an aside, it should be pointed out that the methods of sludge stabilization in present use may actually exacerbate some of these problems while
reducing others. For example, anaerobic digestion may produce toxins from
some of the compounds used as builders for household detergents,[3] and
may also make dewatering more difficult. It is well known that raw primary
sludge is easier to dewater centrifugally than digested sludge.

MEASURING SLUDGE STABILITY

Once a disposal or use method has been selected (perhaps only tenta-
tively), and the characteristics of the sludge in question have been determined, parameters which define acceptable sludge stability must be
selected.

It is common to see any of the following parameters used for measuring
stability:

- volatile solids
- BOD$_5$ of the filtrate
- increase in BOD$_5$ with storage
- oxygen uptake
- gas production
- volatile acids
- pH change during solids
- ATP, DNA and enzymatic activity

Some authors have combined several of these to create an index of stability.[4]

It is important to recognize that *none of these parameters measure stability*, as defined above. They are all *indirect* measures, relying on association. For example, sludges with high volatile solids tend to be odiferous, so volatile solids are commonly used as a parameter for measuring stability. High ATP and DNA levels indicate high concentrations of viable biomass, and this could be indicative of pathogens and may lead to odor production, but ATP and DNA do not measure *stability*, as defined above.

As stated earlier, the problems associated with the use or disposal of raw sludges are: (1) odor, (2) pathogens, (3) toxins, and (4) poor dewaterability.

Odor

The difficulty with odor measurement is of course that each person has different olfactory sensitivity. The qualitative measure of odor (does it smell "good" or "bad") is even more difficult.

Quantitative measurement of odor intensity is possible by using a panel of judges and asking each judge to sniff progressively diluted odoriferous gases until the odor is no longer detected. The number of dilutions needed to attain odor-free air is the *odor number*, a technique borrowed from the water treatment industry.[5]

Two difficulties arise with this technique. First, each judge has different sensitivity, and second, the test cannot readily be conducted in the field since it requires substantial laboratory equipment. A further disadvantage in defining the odor number as the number of dilutions is that the range of odor detection for the human nose is very wide. For example, the detection threshold of H$_2$S is 0.5 ppb, while the concentration of H$_2$S in sewers is about 1000 ppb, and the lethal level is about 2,000,000 ppb.

These objections can be overcome by using a *dynamic-olfactometer* which allows for field measurement, and by defining the odor as

Personalized Specific Odor Number $= \log_{10} C \times S/C_t$

where

    C   = concentration of the odor-containing gases
    C_t = threshold level of these gases
    S   = individual sensitivity

If the flow rate of odor-free air in the olfactometer is Q_1 and the flow rate of odor-containing air is Q_2, then it can be shown that

$$PSON = \log_{10} \frac{(Q_1 + Q_2)S}{Q_2}$$

The sensitivity term, S, is defined as

$$S = \frac{C_{tm}}{C_t}$$

where

    C_{tm} = measured threshold value for the individual doing the sampling, using a gas such as H_2S
    C_t = published and "accepted" threshold value

Thus each investigator would have a sensitivity number which corresponds to his or her individual olfactory efficiency. Commonly S is thought to be about 5.[6]

To date, measurements using this technique have not been published. It is suggested here as a convenient quantitative parameter for one measure of sludge stability.

Pathogens

Typically, *E. coli* are used as indicators of the presence of pathogens, and the vast majority of studies on sludge disinfection use the coliform organisms as an indicator.

It should be recalled that *E. coli* were originally chosen as *indicators* of pollution, i.e., the presence of these organisms indicates the *possibility* of the presence of pathogens. In sludges, we know already that pathogens exist, so that *E. coli* are really not indicating anything. These organisms may not, therefore, be the appropriate microbes for estimating the quantity of pathogens. Salmonellae, for example, may be a much more accurate quantitative measure of pathogens, and it is suggested that an organism (such as salmonellae) which truly represents pathogen concentration be selected and used in sludge work as one measure of stability.

Toxins

Two types of toxins — inorganic and organic — are of importance. Of the inorganic variety, heavy metals seem to be of greatest concern. Their measurement is fairly well standardized using atomic adsorption spectrophotometry or other analytical techniques. Similarly, organic toxins such as PCB or polychlorinated hydrocarbons can be accurately measured.

Dewaterability

The ability of a sludge to yield water and thus concentrate its solids is not directly connected to its disposal, and usually involves only the economics of transport or further processing, such as incineration. Nevertheless, sludge dewaterability is of great concern to most treatment plant operators, and sludges are often treated (stabilized) for the sole purpose of enhancing their dewaterability.

At present, it is possible to apply only one of three mechanisms for dewatering:

1. filtration
2. centrifugation
3. evaporation

Because all three are fundamentally different operations, dewaterability must be measured according to the specific operation used in plant scale.

For filtration, the specific resistance to filtration seems to have emerged as the standard technique, although there is no ASTM or *Standard Methods* procedure yet available.

In centrifugation, it has been shown[7] that the compaction of sludge under centrifugal acceleration seems to be independent of its initial concentration. Centrifugal dewatering requires not only that the solids settle, but that they be sufficiently firm to be able to be moved out of the centrifuge by the screw conveyor. This firmness can be defined using a standard penetrometer, a plastic rod dropped into the compacted sludge.[8] The depth of sludge not penetrated is a measure of the ease of movement out of a full-scale centrifuge.

Although stabilization by digestion is often practiced because the sludge is to be dewatered on sand drying beds, the ability of a sludge to drain and evaporate is seldom measured. Drainage only accounts for a small fraction of the water loss in sand drying beds, and thus the ability of a sludge to yield water should be an important property. Some sludges, for example, will form a heavy crust on top and effectively prevent any evaporation, thus making their dewatering by sand beds impractical. No standard method for evaluating the evaporative potential of sludges has been suggested, although some laboratory devices have been shown to be useful for evaluating the use of chemical conditioners.[9]

STABILIZING OPTIONS

Table 2.1 is a listing of the commonly employed methods of sludge stabilization, along with their corresponding effects on the four sludge stability parameters discussed above. It should be noted that none of the techniques provide complete stabilization, including incineration, which results in a residue high in heavy metals (and a flue gas which contains other heavy metals such as mercury and possible secondary organic toxins such as dioxin). It should be evident that the selection of a stabilization process requires the knowledge of what the ultimate disposal will involve and what

Table 2.1.   Effectiveness of Sludge Stabilization Processes[a]

Method	Effectiveness			
	Odor Reduction	Pathogen Destruction	Toxin Removal	Dewaterability Enhancement
Anaerobic digestion	+	+	+[b]	+[c]
Aerobic digestion	+	+	0	−[c]
Lime treatment (Ca(OH)$_2$)	+	+	+[d]	+
Quicklime (CaO)	+	+ +	+ +[d]	+
Composting	+	+	0	NA[e]
Lagoons	0	+	0	+[c]
Chlorination	+	+ +	0	+
Irradiation	0	+ +	0	0
Pasteurization	−	+ +	0	+
Drying	+ +	+ +	0	NA[e]
Incineration	+ +	+ +	0	NA[e]

[a] − = negative effect, 0 = not very effective, + = somewhat effective, + + = very effective.
[b] Supernatant removal.
[c] On sand beds only.
[d] Reduction of soluble metal.
[e] NA, not applicable.

sludge characteristics are required. For example, disposal on strip mine land would require dewaterability since the transport distances are great. If the land is far removed from civilization, the wisdom of pathogen destruction or odor reduction is questionable. On the other hand, if the sludge is to be given away to private gardeners, both odor and pathogens as well as toxins would be of great importance, and the stabilization techniques selected should reflect these requirements.

DESCRIPTION OF SLUDGE STABILIZATION OPTIONS

Anaerobic Digestion

For over two hundred years it has been known that putrescible material, left to stand in an enclosure, will degrade, producing a humus-like substance. In the early 1900s Karl Imhoff was the first to apply this idea to wastewater treatment by designing settling tanks which had space for anaerobic digestion of the settled solids. In the 1920s, the idea of *separate* digestion became popular, and the anaerobic digester as we now know it was developed. With only minor modifications, the digester of today is identical to the units installed over 50 years ago.

Process

The anaerobic digestion process depends on the action of a number of microorganisms, generally classified into the *acid formers* and the *methane formers*. The process can be characterized as the following:

raw sludge organic solids $\longrightarrow$ extracellular enzymes $\longrightarrow$ dissolved organic solids which can be utilized by the acid formers, producing $\longrightarrow$ organic acids $\longrightarrow$ methane formers, producing $\longrightarrow$ methane, carbon dioxide, etc.

The methane formers, a diverse group of strict anaerobes, are extremely sensitive to changes in environment, and their welfare is one of the primary concerns of wastewater treatment. There seem to be only certain ranges of temperature, alkalinity, pH, etc., that will provide for adequate methane former survival. They are also slow growing, and thus the solids retention time in an anaerobic digester must be sufficiently great not to wash out these microorganisms. The solids retention time, calculated as

$$\Theta = \frac{\text{solids in the digester, lb of dry solids}}{\text{pounds per day of dry solids added to the digester}}$$

must be at least 10 days in order for adequate digestion to occur. Commonly solids detention times are from 20 to 60 days (but this is highly dependent on mixing, as discussed later).

The pH of an anaerobic digester must remain above 6.0 for adequate digestion to occur, but this is the last of the operating parameters to worry about. Low alkalinity is a precursor of problems with pH and is the parameter which must be controlled in operation. Figure 2.2 shows the normal limits of alkalinity as plotted against the fraction of CO_2 in the gas produced. This is the second most sensitive parameter indicating the health of the digester. When the fraction of CO_2 falls below about 25% or increases to above 45%, the digester is headed for trouble.

Temperature is a major design criterion, in that there seem to be two groups of organisms which perform well at fairly narrow temperature ranges—between 27 and 43°C, called the mesophilic range, and 45 and 65°C, called the thermophilic range. Although the latter seem to be hardier and considerably faster growing (hence producing more rapid reduction in volatile solids) the temperature is sufficiently high to make the use of thermophilic digestion in many parts of the United States unrealistic.

Hardware

The typical anaerobic digestion system used in the United States today is the combination of primary and secondary digesters, pictured in Figure 2.3. The primary digester is mixed and heated, while the secondary digester is used primarily for storage of the digested sludge and the gas. Typically the secondary digester has a floating cover to accommodate the storage of the gas.

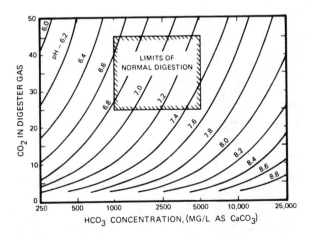

Figure 2.2. Relationship between CO_2 in the digester gas and bicarbonate alkalinity. (*Source*: Reference 10.)

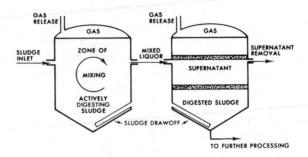

Figure 2.3. Typical two-stage anaerobic digesters.

Design

Design criteria for typical digesters are shown in Table 2.2.

Probably the most difficult problem in design is mixing of the anaerobic digesters. Present methods do not provide adequate mixing, as testified by numerous studies which, using dye tracer studies, have found that only about 25 to 50% of the digester volume is actually being used for active digestion, the remaining being dead space filled up by inert solids or scum. Present mixing methods seem to be inadequate and better hardware is

Table 2.2. Typical Digester Design Criteria

Parameter	
Solids retention time (days)	10 to 20
Solids loading (lb VSS/ft^3/day)	0.15 to 0.40
Volume criteria (ft^3/capita)	
Primary sludge	1–1/3 to 2
Primary sludge + trickling filter sludge	2–2/3 to 3–1/3
Primary sludge + waste activated sludge	2–2/3 to 4
Combined primary + waste biological sludge feed concentration, (% solids, dry basis)	4 to 6
Digester underflow concentration, (5 solids, dry basis)	4 to 6

Source: Reference 11.

needed. Of the available technology, the two-stage slow-speed turbine mixer seems to be the best, and is highly recommended in the absence of better alternatives.[12]

Operation

Although from an operational standpoint, the secondary digester provides storage for emergencies, and a method of returning some of the liquid fraction to the plant, it is in fact a wasteful and detrimental part of a wastewater treatment plant as far as sludge management is concerned. Supernatant (or what some operators laughingly call their supernatant) is often so high in fine particle suspended solids that its recirculation into the treatment plant can cause serious damage to treatment efficiency. The storage of sludge from the primary digester results in a sludge that is less amenable to dewatering, and requires more chemicals for conditioning, than the sludge from the primary digester.

Performance

Typical digesters will reduce the volatile solids by 40 to 45%, but this is not an adequate measure of stabilization, as discussed above.

There are few data available on the odor reduction in digestion, but there similarly is no argument that anaerobic digestion does indeed make the sludge considerably more pleasant to be around.

Reduction of pathogenic organisms is another important parameter describing stabilization, and the process does reduce pathogens substantially, as shown in Table 2.3. It should be emphasized that anaerobic digestion does not produce sterile sludge, and does not qualify for EPA's criteria as a process that "further reduces pathogens."

Aerobic Digestion

An alternative to anaerobic digestion is aerobically accomplished by allowing sufficient contact time between aerobic organisms and the volatile organic matter. In outward appearance, the aerobic digester differs little from an aeration basin used in the activated sludge system. In operation, however, it is quite different, since the objectives are not the same. In activated sludge aeration, the purpose is to destroy the dissolved organic matter, as measured by the biochemical oxygen demand, using microorganisms which convert these organics into biomass, whereas in the aerobic digestion operation the objective is to destroy volatile *suspended* organics, that is, to destroy the very organisms that were grown for a useful purpose in the activated sludge system.

Table 2.3.   Bacterial Survival in Digestion

Bacteria	Digestion Period (days)	Removal (%)	Remarks
Endamoeba hystolytica	12	< 100	Greatly reduced populations at 68° F
Salmonella typhosa	20	92	85% reduction in 6 days detention
Tubercle bacilli	35	85	Digestion cannot be relied upon for complete destruction
Escherichia coli	49	< 100	Greatly reduced populations at 99° F; about the same reduction in 14 days at 72° F

Source: Reference 13.

Process

Very simply, the process involves long-term aeration and storage of the biomass so that further microbial action will reduce the total volatile solids. In treatment plants where activated sludge is already employed, the digestion of waste activated sludge by aerobic means seems to make a great deal of economic and practical sense. The process has enjoyed less success in the digestion of raw primary sludges, however.

Hardware

Aeration can be either by diffused aeration or mechanical surface aeration, or any other aeration option. In treatment plants where diffused air is used in the activated sludge system, it is economical to simply bleed off some of the air for the digester. Figure 2.4 shows a typical aerobic digester operation, although in many smaller plants the settling tank is omitted and the sludge is dewatered and/or disposed of directly from the digester.

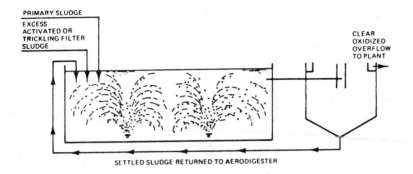

Figure 2.4. Aerobic digestion.

Design

Table 2.4 shows some design data for aerobic digesters. Typically, these systems require at least 15 days of aeration in order to achieve sufficient volatile solids reduction.

Operation

With 15 days of aeration as the hydraulic detention time, the process will perform adequately if biological sludges are digested. With significant amounts of primary sludges (a situation which should not arise) the required detention time is greater. Sludge age is typically around 50 days, with loadings around 0.01 lb volatile solids/ft^3/day.

Performance

It is not unusual to encounter volatile solids reductions as high as 50 to 60%, although a 40% reduction is also commonly attained. If aerobic digesters have supernatant return, these supernatants tend to be clear and low in BOD, thus not creating the problems the anaerobic supernatants impose on the plant.

Stabilization with Slaked Lime

Slaked lime, or $Ca(OH)_2$, will raise the pH of sludge to about 12 and thereby achieve substantial reduction in pathogenic organisms. In addition, lime is an effective additive when used in conjunction with aluminum or iron salts, as a sludge conditioner prior to dewatering. In order for Al^{+3} or Fe^{+3} to be effective, they must have a sludge with a sufficiently high pH,

Table 2.4. Aerobic Digestion Design Parameters

Parameter	Value	Remarks
Solids retention time (days)	10–15[a]	Depending on temperature, type of sludge, etc.
Solids retention time (days)	15–20[b]	
Volume allowance (ft³/capita)	3–4	
VSS loading (lb/ft³/day)	0.024–0.14	Depending on temperature, type of sludge, etc.
Air requirements		
diffuser system (cfm/1000 ft³)	20–35[a]	Enough to keep the solids in suspension and maintain a DO between 1 and 2 mg/l.
diffuser system (cfm/1000 ft³)	>60[b]	
Mechanical system (hp/1000 ft³)	1.0–1.25	This level is governed by mixing requirements. Most mechanical aerators in aerobic digesters require bottom mixers for solids concentration greater than 8000 mg/l, especially if deep tanks (>12 ft) are used.
Minimum DO (mg/l)	1.0–2.0	
Temperature (°C)	>15	If sludge temperatures are lower than 15°C, additional detention time should be provided so that digestion will occur at the lower biological reaction rates.
VSS reduction (%)	35–50	
Tank design		Aerobic digestion tanks are open and generally require no special heat transfer equipment or insulation. For small treatment systems (0.1 mgd), the tank design should be flexible enough so that the digester tank can also

Table 2.4., continued

Parameter	Value	Remarks
		act as a sludge thickening unit. If thickening is to be utilized in the aeration tank, sock type diffusers should be used to minimize clogging.
Power requirement (BHP/10,000) Population equivalent	8–10	

Source: Reference 13.
[a]Excess activated sludge alone.
[b]Primary and excess activated sludge, or primary sludge alone.

and lime is normally employed for this purpose. An additional advantage of lime is that its use greatly reduces the odor problem.

The most difficult problem with the use of slaked lime is that, in time, the effect of lime wears off, and as the pH is depressed, the odiferous characteristics of the sludge return. Thus lime is not really a stabilizing agent for permanent odor control, but it is useful in the destruction of pathogens.

Stabilization with Quicklime

Quicklime, or CaO, reacts with the water in the sludge as follows:

$$CaO + H_2O \longrightarrow Ca(OH)_2 + heat$$

This produces two beneficial effects. First, some of the water is chemically bound to the calcium, thus effectively dewatering the sludge. Secondly, the combination of the heat and pH produces an effective mode of pathogen destruction. Present experience indicates that if lime is added at about 40% by weight to already dewatered sludge (25%) solids, a crumbly, disinfected product is obtained which might have significant market value. The quicklime process is described in greater detail in the case study portion of this book, under the title of "lime encapsulation."

Drying

Drying is the process of using thermal energy to evaporate water and produce a product which is essentially disinfected, odor free, and very low in moisture (about 5 to 10% moisture). In some treatment plants, such as the one in Largo, Florida, drying is used to produce a marketable fertilizer and soil conditioner.

There is no doubt that drying, either by flash dryers or rotary kilns, produces a very fine product. The problem arises in the economics and the eventual marketing of the dried sludge. Typically, municipalities have been notoriously inept in the marketing of the finished product, and thus many of the drying systems designed over the years have failed because of inability to sell the sludge to farmers or fertilizer manufacturers. In addition, there have been several serious accidents involving sludge drying due to the production of a very fine organic aerosol which is highly explosive.

Often the economic success or failure of the drying operation depends on the effectiveness of dewatering. A few percentage points in cake solids concentration translates into considerable fuel savings in the drying process.

Incineration

Incineration is the most effective means of sludge stabilization. By fully oxidizing the various organics, all odor is eliminated and all pathogens are destroyed. The composition of the sludge is an important variable in incineration, not only in its calorific value (BTU/lb solids, or in SI units, kJ/kg dry solids), but also in the concentration of heavy metals and refractory and toxic organic compounds. These contaminants can be emitted with the offgases, and may create air quality problems. In addition, some organics, such as some members of the dioxin family, may actually be *produced* during incineration and thus present a serious health concern.

Process

The incineration process can be represented schematically as in Figure 2.5. The basic operation is thermal oxidation, at temperatures sufficiently high to cause the thermal degradation of organic compounds, producing an inert ash and resulting in about 90% volume reduction of the sludge.

Hardware

Although there are any number of incinerator types available, the only two that seem to have found use in sludge incineration are the multiple-hearth unit and the fluid bed unit. These are shown schematically in Figures 2.6 and 2.7.

In the multiple-hearth unit, sludge enters in the top of a series of hearths, and moves down by gravity, from one hearth to the next, assisted by rotat-

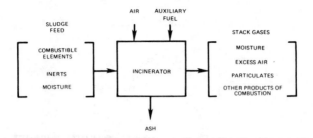

Figure 2.5.   Schematic representing the basic processes in sludge incineration.

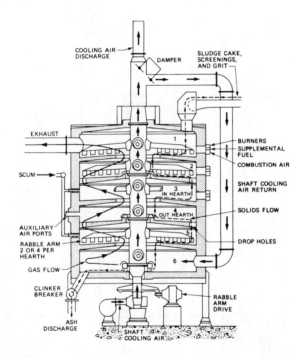

Figure 2.6.   Multiple-hearth incinerator.

ing rabble arms. The topmost hearth is used for drying since the hot gases emanating from the middle combustion zone flow past the wet sludge. The lowest hearths are used for cooling the ash, since the cold air enters there and is in turn heated so it can provide the oxygen for the combustion in the middle hearths. Figure 2.8 shows the concept of the concurrent multiple-hearth process.

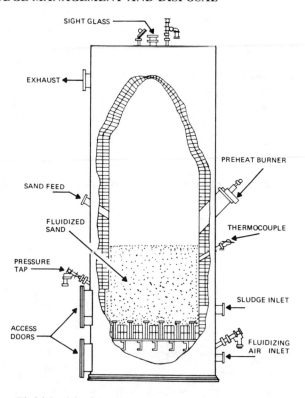

Figure 2.7.   Fluid bed incinerator.

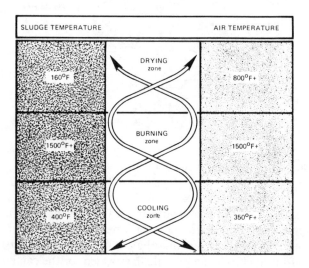

Figure 2.8.   Process zones in a multiple-hearth incinerator.

The fluid bed system uses a "thermal fly wheel"; sand which is heated and kept "boiling" by means of air blown in underneath. Sludge is introduced into this hot, boiling sand and oxidized. The temperature in the fluid bed incinerator is usually about 800°C.

Design

Sludge incinerators are designed much like other incinerators — on the basis of the BTU or calorific value of sludge. Typical values of common sludges are shown in Table 2.5.

Operation

The multiple-hearth incinerator is so designed as to offer many points of control, including the sludge feed rate, bottom air rate and rabble arm rate. For many years, sewage treatment plant operators just were not trained to operate these units and performance was almost uniformly bad. Only recently, with help from the EPA, have some of the multiple-hearth units enjoyed considerable economic success.

Performance

When operating properly, sludge incinerators produce a stable inert ash which is commonly low enough in heavy metals to avoid being classified a hazardous waste. Emissions, while a continuing concern, have not been shown to date to be a public health hazard. The major problem with performance is the dewatering of the sludge to sufficiently high solids concentration to reduce the amount of auxiliary fuel required.

Table 2.5.   Representative Heating Values of Some Sludges

Material	Combustibles (%)	Higher heating value (Btu/lb of dry solids)[a]
Grease and scum	88	16,700
Raw wastewater solids	74	10,300
Fine screenings	86	9,000
Ground garbage	85	8,200
Digested sludge	60	5,300
Chemical precipitated solids	57	7,500
Grit	33	4,000

Source: Reference 14.
[a]1 Btu/lb = 2324 MJ/kg.

Composting

Composting is the aerobic decomposition of organics, with the process carried on while the sludge is in a solid or semisolid state, with oxygen being provided by either agitating the sludge, or blowing air through the sludge mass. With the latter process, bulking agents are necessary to provide the interstices for the air to flow through. Because composting is aerobic, it is an exothermic operation and the heat created is effectively used for the destruction of pathogenic organisms.

Process

In years past, wet sludge (not dewatered) has been used as a source of nutrients and moisture in the composting of shredded municipal solid waste. This process no longer seems to be used in the United States, and thus is not discussed further in this book. Instead, it is assumed that dewatered sludge is to be composted, either before or after digestion.

Two distinctly different operations are used for sludge composting. In one, often termed "closed vessel composting," the sludge is contained in a closed vessel such as a rotary kiln, and tumbled or otherwise agitated to provide oxygen for the microorganisms. The second option requires the use of bulking agents, such as wood chips, to form passages for the movement of air through the static pile, and the sludge is only occasionally agitated, or air is provided by means of air blowers.

Hardware

Figures 2.9 and 2.10 show examples of these two composting options.

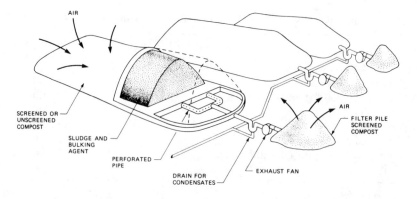

Figure 2.9. Static aerated pile composting.

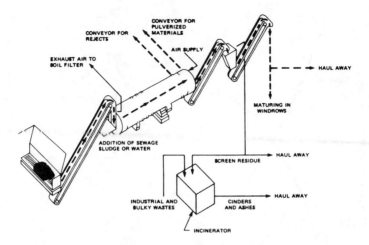

Figure 2.10.   Typical enclosed-vessel composting plant.

Design

The objective of composting is to produce a disinfected and odor-free product which can be used in agriculture or in other land treatment applications. Thus the proof of any composting operation is in the product.

In designing a composting operation, perhaps the most critical parameter is the availability of nutrients. Fortunately, wastewater sludges have sufficient quantities of nitrogen to allow the reaction to occur without the use of inorganic nutrients. Typically, the C/N ratio is about 1/16; 1/20 is required for composting.

Operation

The most important operational variable in composting is moisture control. Should moisture fall below about 40%, the microorganisms cannot operate, while a moisture concentration above 60% will cause the formation of anaerobic clumps of sludge. Thus it is necessary, whatever hardware or system is used, to provide for just the proper moisture.

Other important variables such as pH are not necessary to control, since the process will adjust itself to the optimal pH.

Performance

Figure 2.11 shows the process of pathogen destruction in a nonmechanical composting operation. Note that the temperature attained is over 60°C,

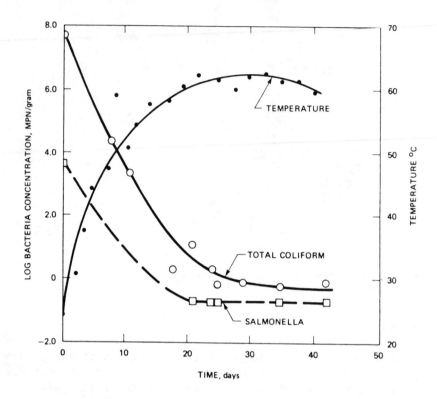

Figure 2.11. Destruction of pathogens in a static pile composting operation. MPN, most probable number (*Source*: Reference 15.)

which is high enough to result in substantial disinfection, but it is not pasteurizing temperature, and thus some pathogenic organisms, such as the ova of parasites, have been shown to survive composting.[15]

Disinfection by High-Energy Radiation

The use of high-energy radiation for wastewater sludge disinfection has been shown to be effective in the destruction of pathogens. Two types of radiation have been employed—beta and gamma. Although effective, it seems unlikely that the cost of such high technology and potentially dangerous systems will soon find wide use.

REFERENCES

1. "Criteria for Classification of Solid Waste Disposal Facilities and Practices" *Federal Register*, 13 September 1979, part IX. Washington, DC.
2. "Final Report of the Management Committee" Commission of the European Communities, COST Project 68, Sewage Sludge Proceedings, Brussels, 1975.
3. Schaffner, C., P. H. Brunner, and W. Giger. "4-Nonylphenol, a Highly Concentrated Degradation Product of Nonionic Surfactants in Sewage Sludge," unpublished report, EAWAG, Dubendorf, Switzerland (1982).
4. Colin, F. "Methodes d'Evaluation de la Stabilite Biologique des Boues Residuaires" Report No. NPSP 28, COST 68 Working Party 1, Commission of the European Communities, Brussels, (1980).
5. Eikum, A. S. "Study of Processes in View of Preventing Odours," paper presented at the 3rd Symposium on Sludge Management, COST 68 Project, Brighton, England (1983).
6. Koe, C. C. L., and D. K. Brady. "Quantification of Sewage Odours," Research Report No. CE40, University of Queensland, St. Lucia, Australia, 1983.
7. Vesilind, P. A., and G. Zhang. "Technique for Estimating Sludge Compactibility in Centrifugal Dewatering," unpublished report (Durham, NC: Duke University, 1984).
8. Vesilind, P. A. "Estimation of Sludge Centrifuge Performance," *J. San. Eng. Div.*, ASCE, 96(SA3) (1970).
9. Davis, H. and P. A. Vesilind. Unpublished Report (Durham, North Carolina: Duke University, 1982).
10. McCarty, P. "Anaerobic Waste Treatment Fundamentals" *Public Works*, 95:107 (1964).
11. Stanley Consultants, Inc. Quoted in Process Design Manual for Sludge Treatment and Disposal EPA 625/1-74-006 (1974).
12. Reardon, D. J., and W. F. Owen. "Fundamentals and Recent Developments in Digester Mixing," paper presented at the 54th Annual California Water Pollution Control Conference 1982.
13. *Process Design Manual for Sludge Treatment and Disposal*, EPA 625/1-74-006, 1974.
14. Owen, M. B. "Sludge Incineration" *J. San. Eng. Div.*, Am. Soc. Civ. Eng., Feb., 1957.
15. Burge, W. D. "Occurrence of Pathogens and Microbial Allergens in the Sewage Composting Environment," National Conference of Composters of Municipal Sludges, Rockville, MD, 1977.

CHAPTER 3

SLUDGE DEWATERING TECHNOLOGY

INTRODUCTION

Solid/liquid separation techniques presently available use one or a combination of only three basic principles:

1. The solids are more (or less) dense than the surrounding liquid.
2. The solids are larger in physical size than the liquid molecules.
3. The solids will not volatilize when the liquid is evaporated.

These three principles are illustrated in Figure 3.1.

Table 3.1 is a listing of a number of commercially available solid/liquid separation techniques, and the basic principles on which they are based. Not all of these are equally applicable to wastewater sludges, and only some of the more commonly used techniques are discussed below.

SLUDGE DEWATERING OPTIONS

Gravity Thickening

Process

Most (but not all) wastewater sludges have particles which are more dense than the surrounding liquid, and these particles will settle under quiescent conditions. Thickening is used to remove the free water which might be carried with the sludge.

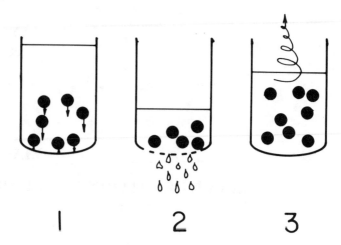

Figure 3.1. Three basic principles used in solid/liquid separation: (1) solids are more dense, (2) solids are bigger, and (3) solids will not volatilize.

Table 3.1.   Techniques for Achieving Solid/Liquid Separation for Wastewater Sludges

General Unit Operation	Specific Equipment	Basic Principle on Which Based
Gravity thickening	Thickening tank	1
	Tube thickener	1
	Lagoon	1
Flotation	Natural flotation	1
	Dissolved air flotation	1
Centrifugation	Solid basket	1
	Disc	1
	Solid bowl	1
	Perforated bowl	1 & 2
	Hydrocyclone	1
Filtration	Vacuum filter	2
	Filter press	2
	Belt filter	2
	Gravity filter	2
Heat treatment	Rotary dryer	3
	Flash dryer	3
	Sand beds	2 & 3

Hardware

The ordinary gravity thickener resembles in many respects a settling tank. As shown in Figure 3.2, the sludge enters the tank through the central core, is distributed radially, and exits through a sludge trough at the bottom. The clear water is discharged over a peripheral weir. The scraper mechanism assists in moving the bottom sludge toward the trough and prevents the formation of "rat-holing," the formation of a water cone during bottom sludge pumping.

Some thickeners have vertical picket fence posts attached to the rotating frame. These rakes are thought to assist in the consolidation process. Actually, it has been proven conclusively that stirring (at even slow speeds) is detrimental to thickening.[1,2] Only in cases where the sludge is biologically active and forms gas are the pickets of value, since they promote the removal of the gas.

A variation of the ordinary gravity thickener is the tube settler, in which a large surface area is created by placing plates or tubes at an angle into the tank. This idea, although old and oft patented, significantly promotes the thickening of some slurries. In order to be removed from the main body of surrounding liquid, the solids have to drop only a short distance and then simply slide down the slanted plate or tube. The trapped water, meanwhile, can travel to the top on the underside of the plates. Maintenance problems have been known to plague some of these systems, but they nevertheless have found acceptance in the thickening of some sludges (as alum sludge) from water treatment plants.

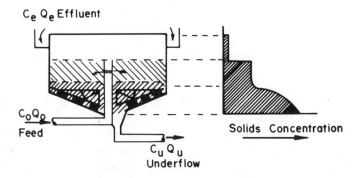

Figure 3.2. Gravity thickener cross section. The graph shows typical solids concentrations in a continuous thickener.

A third thickening technique is the lagoon. In this case, a large hole in the ground is filled up with the sludge, the sludge is allowed to consolidate, and the liquid is periodically decanted. More sludge is added as space allows. Eventually, of course, lagoons fill up and must be covered over or dredged out. Although lagoons may seem to be expedient means of sludge management for the short term, they invariably will result in serious problems and high disposal costs in the end.

Design

In all gravitational thickening applications, the processes are sized on the basis of how well the sludge settles. As the settleability improves, the required areas and volumes decrease.

Sludge settleability has historically been evaluated on the basis of tests in a 1-liter cylinder. The classical sludge volume index (SVI), for example, is defined as the volume occupied by 1 g of sludge after 30 min settling in a 1-liter graduated cylinder. More recently, such tests have been used to develop design criteria for thickeners.

The prevailing sludge thickening design procedure allows for the calculation of the required surface area in order that a sufficient quantity of solids can be transmitted through the tank per unit time. This is referred to as "solids flux," and is expressed in terms of the mass of solids moved through a unit area in a given time (e.g., kg solids/hr/m^3 surface area).

The required area for thickening can be estimated by calculating the critical solids flux — that flux which will limit the operation of the thickener. A series of batch thickening tests is conducted using sludge at different solids concentrations. The height of the sludge/liquid interface is recorded with time, and the results plotted, as in Figure 3.3. The batch flux is defined as the settling velocity times the solids concentration. Note that the units are consistent;

$$G_B = vC = \frac{m}{hr} \times \frac{kg}{m^3} = kg/hr/m^2 = \text{batch flux}$$

where G_B is the batch flux, v and C are the settling velocity and solids concentration, respectively.

In a continuous thickener, the total flux consists not only of the solids settling through a given area, but also includes the movement of the solids downward due to the removal of the underflow. Consider for a moment a thickener which has solids that do not settle at all (i.e., same density as the liquid). Obviously, this slurry will have no batch flux (v = 0, hence $G_B = 0$ for all C). But placed in a continuous thickener, a downward flux of solids

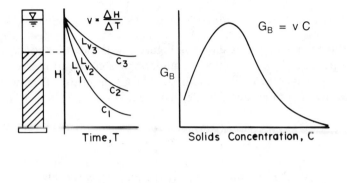

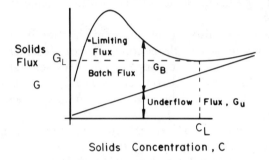

Figure 3.3. Calculation of limiting flux. The velocity of the sludge/liquid interface is first measured for different concentrations. By multiplying the velocity by the corresponding solids concentration, the flux is calculated and plotted against the concentration. The minimum flux is determined by superimposing the batch flux on the underflow flux and determining the low point in the composite curve.

will exist as long as there is extraction of the underflow. This flux is known as the underflow flux, and defined as

$$G_u = \left(\frac{Q_u}{A}\right) C$$

where G_u is the underflow flux, Q_u is the flow rate of the underflow and A is the thickener area

This underflow flux can then be plotted on a graph of flux vs concentration and will be a straight line for a given Q_u and A.

The total flux is thus

$$G = G_B + G_u$$

and this can be plotted as shown in Figure 3.3. If the solids concentration of the influent is C_o and the desired underflow concentration is C_u, the total flux curve passes through a minimum. This is the critical flux, which determines the minimum area for the thickener, since

$$G_L = \frac{Q_o\, C_o}{A_{min}}$$

where A_{min} is the minimum area. If the loading is too high, the thickener cannot handle the solids, and some of them will be discharged over the effluent weir.

Tests with cylinders can also provide an approximation of the possible benefit of gravity thickening, i.e., the concentration of the underflow. This is obtained by allowing the sludge to settle to its final volume and taking the ratio of the heights, so that $C_u = C_o(H_o/H_f)$, where C_o is the initial solids concentration, H_o is the initial height of the sludge in the cylinder, and H_f is the final compacted height.

A complete description of the thickener theory may be found elsewhere.[1] The important aspect of this analysis is that thickeners can only be loaded to a limit, at which point they will not accept additional solids.

It is necessary to also caution against the use of small (e.g., 1-liter) cylinders for the design of large (and expensive) thickeners. For reasons not yet adequately understood, small cylinders exert a scale effect on sludge settling. As shown in Figure 3.4, cylinders shorter than 3 ft tend to produce

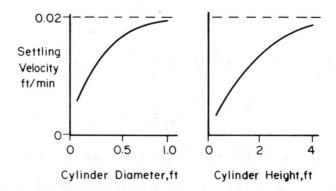

Figure 3.4. Effect of initial depth and cylinder diameter on settling velocity for a typical activated sludge at 2000 mg/1. [Reference 3]

lower settling velocities,[3] and cylinders less than 12 in. in diameter tend to increase or decrease the velocity, depending on the sludge and the concentration.[4] Additional problems with batch testing include methods of filling the cylinder and other laboratory artifacts.

Operation

Gravitational thickeners are used both in batch and continuous operation. Good design dictates that only one type of sludge be used — that different sludges not be mixed in the thickener.

Performance

With waste activated sludge, underflow solids concentrations range from 1.5 to 4.0% solids. Raw primary sludges have been thickened to as much as 8% solids.

Flotation Thickening

Process

Wastewater sludges sometimes tend to be very light, and thus settle and compact poorly under the force of gravity. For example, waste activated sludge flocs have densities of less than 1.08 (density, or specific gravity, of water = 1.0). Accordingly, it is sometimes easier to achieve separation of solid flocs from liquid by promoting their flotation.

Hardware

Flotation thickeners have been used for a number of years in the thickening of waste activated sludge. As shown in Figure 3.5, the flotation unit requires that air be dissolved under pressure and then mixed with the incoming slurry. As the pressure is released, the air bubbles attach themselves to the sludge particles, increase their buoyancy, and cause them to travel to the top, where the thickened solids are skimmed off.

In almost all cases, the sludges thickened by flotation are conditioned with chemicals, such as organic polyelectrolytes. Such "polymers" tend to agglomerate the particles and create discrete flocs which can then be buoyed upward. Tiny particles, although light, require too long to rise and would not be removed.

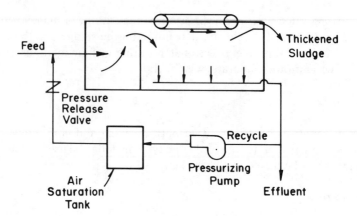

Figure 3.5. Dissolved air flotation thickener. Air is dissolved under pressure and then released. The small bubbles attach themselves to the floc particles and carry them to the top.

Design

The amenability of a sludge to be thickened by flotation can be tested by a simple setup illustrated in Figure 3.6. As the compressed air in water is released into the sludge, the sludge is mixed, and the small bubbles are attached to the solids, which begin to rise. The rate of the sludge/liquid interface is recorded, and its velocity is exactly analogous to the settling velocity in a batch thickening test. The required thickening area can be calculated in a like manner.

The extra variables present are, of course, the amount of air and its pressure. The effect of the air is best described on the basis of the ratio of the air to the solids (weight to weight). The effect of the air-to-solids ratio is described in Figure 3.7. Note that as the amount of air is increased for a given solid, the recovery and the thickened solids concentration both achieve practical limits and there is nothing to be gained by further addition of air to the thickener. The most efficient operating point can be ascertained by conducting a series of laboratory tests at various air/solid ratios.

It should be mentioned that the apparatus pictured in Figure 3.6 is subject to the same potential laboratory artifacts as discussed under gravitational thickening. Further, the laboratory system does not accurately reflect the actual operation of a dissolved air flotation thickener, in that the air is mixed with the sludge as a batch operation all at once, whereas in a full-scale thickener the water containing the dissolved air is continuously mixed with the sludge and then the pressure is released on the mixture. This

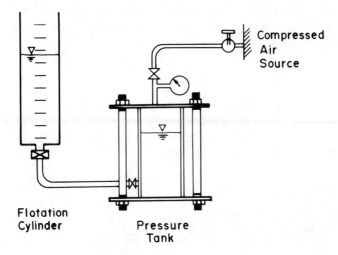

Figure 3.6. Flotation apparatus. Air is dissolved in the cylinder under pressure and mixed with the sludge in the flotation cylinder. The rise of the sludge/liquid interface is recorded with time.

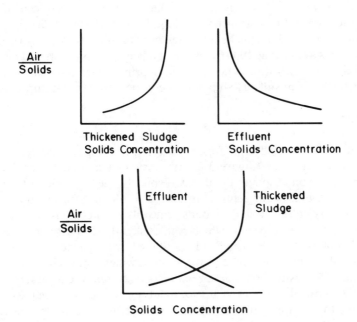

Figure 3.7. The air-to-solids ratio affects both the thickened solids and the recovery of solids in a dissolved air flotation system.

operation can be simulated by simply blending the sludge and water into the flotation cylinder at the required ratios. Apparently, more accurate results can be obtained with such a modification.[5]

Operation

Flotation thickeners are used almost exclusively for light, biological sludges. Raw primary sludges are usually too heavy to float well. Chemical conditioning is used almost without exception, even though the process is often marketed as not requiring chemicals.

Performance

Waste activated sludges can be thickened to up to 6% solids, with chemical conditioning, although 4% is more common.

Centrifuging

Process

As with gravitational and flotation thickening, the process of centrifugation is the separation of solids from liquid on the basis of density. The difference is that while thickening uses gravitational force, centrifuging multiplies this as much as 1000 times gravity. Nevertheless, it should be kept in mind that centrifugation is no different in principle from thickening, and most of the problems of scale-up and operation apply to both processes.

Hardware

The most widely used centrifuge for wastewater sludge treatment is the solid bowl or decanter (Figure 3.8). It consists of a solid, tapered bowl which is rotated on its longitudinal axis. Typically, the large end of the bowl has holes which allow the centrate to escape, while the tapered end, called the beach, has holes for sludge solids removal. As the solids settle to the inside wall, a screw conveyor, which rotates slightly slower than the bowl, moves the solids to the open end and out of the bowl. The feed is introduced by means of a central feed pipe (not rotating), which sprays the sludge into the machine. It has for some time been recognized that the introduction of the feed is a critical component of successful centrifugation. The feed tends to splash into the rapidly rotating sludge and destroy some of the separation that has already occurred. Further, it is necessary to accelerate the feed from zero to perhaps 4000 rpm in a very short time. The typical hydraulic residence time in a solid bowl centrifuge is about 20 sec. The recognition of

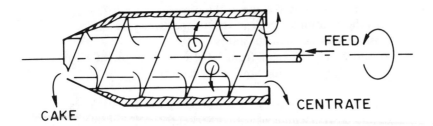

Figure 3.8. Solid bowl centrifuge. The sludge is fed into the rotating bowl, the liquid is skimmed off the top, and the solids are moved by the screw conveyor up to the beach and out of the bowl.

this problem has prompted the development of newer machines which run at slower speeds and which introduce the sludge to the bowl by accelerating it first with an inner cone. These machines have found wide acceptance in the dewatering of light and fluffy sludges, such as activated sludge, alum sludge, and other chemical sludges.

Design

Since centrifuges are basically highly efficient thickening tanks, it is not surprising that the sizing of centrifuges is based on surface area and hydraulic retention time. The most widely used technique for sizing centrifuges is the Sigma concept first developed in 1952 as a means of scaling up geometrically similar centrifuges.[6] Assuming that the particles in the centrifuge settle in the laminar flow regime, are unhindered in settling by their neighbors, and accelerate immediately to rotational speed upon introduction into the centrifuge (all untenable assumptions), it can be shown that the flow rates at equal performance* for two geometrically similar solid bowl machines can be related as

$$\frac{Q_1}{Q_2} = \frac{\Sigma_1}{\Sigma_2}$$

where

$$\Sigma = \frac{V\omega^2}{g \ln (r_2/r_1)}$$

---

*Some authors call this the centrifuge capacity. In truth, a centrifuge does not have a liquid handling capacity. At higher flow rates, performance simply suffers.

where V is the hydraulic volume in the bowl

ω is the rotational speed in radians per second

g is the gravitational constant

r_1 and r_2 are the radii from bowl centerline to the surface of the sludge and inside bowl wall, respectively

Sigma has different formulations for other types of centrifuges.

For a solid bowl machine, as with other centrifuges, another limitation on performance is the solids loading. If a machine is loaded with solids at a rate higher than the conveyor can move them out, there is a solids buildup in the bowl, reducing the hydraulic volume, and adversely affecting performance. The solids loadings can thus also limit a centrifuge, and the relationship between any two geometrically similar machines can be related using the Beta concept,[7] where

$$\frac{Q_1}{Q_2} = \frac{\beta_1}{\beta_2}$$

where β is defined as

$$\beta = \triangle \omega \, S \, N \, \pi \, Dz$$

where $\triangle \omega$ = speed difference between bowl and conveyor

S = pitch of conveyor

N = number of leads

D = diameter of bowl

z = pool depth

It is important to recognize that the performance of a centrifuge can be limited by liquid as well as solids loading, as shown in Figure 3.9. These data, obtained using a calcium carbonate slurry, show that at a given flow rate, performance could not be maintained when the solids concentration was high. At that point, the machine performance became solids limited.

When data from a smaller, geometrically similar machine are not available, it is necessary to estimate the required size on the basis of laboratory tests. The most common test is to centrifuge the sludge in a test tube desktop machine, judge the clarity of the centrate, and estimate the cake consistency by poking at the sludge cake with a glass rod. Both characteristics are important, since the particles must first settle out, and then they must be conveyed by the screw. A well settling sludge that produces a very soft cake is not a good candidate for solid bowl centrifugation.

These tests can be quantified to allow for the comparison of different sludges and to detect changes in a sludge with time.

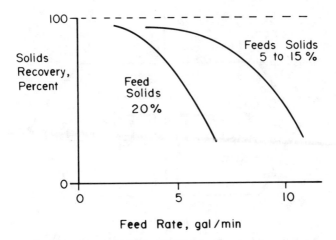

Figure 3.9. Solids recovery of a dilute and concentrated calcium carbonate slurry. Note that for the concentrated slurry, the performance is solids limited.

The settleability of a sludge in a test tube can be measured using a strobe light synchronized with the spinning test tubes, as shown in Figure 3.10. If the shield holding the test tube has a hole in it, it is possible to observe the sludge/liquid interface, and determine its height with time. This information, of course, yields the velocity of sludge settling at a given centrifugal acceleration and solids concentration. The analysis described earlier for estimating the required surface area of a thickener can be applied equally well here for the calculation of centrifuge surface area.[1]

Further, it is possible to use this test to develop a measure of sludge settling characteristics under a centrifugal acceleration.[8] The "settleability coefficient" has been defined as

$$S = \frac{v}{\omega^2 r}$$

where v is the velocity of the interface at some distance r from the centerline, rotating with a centrifugal speed of ω radians per second. The value of S seems to be independent of the centrifugal force imposed, and thus is a true measurement of how well the sludge settles.

The second characteristic, the sludge firmness or body, can be evaluated with a penetrometer, shown in Figure 3.11. A metal or plastic rod is dropped into the sludge cake and its penetration measured. A sludge which will move readily in a scroll centrifuge cannot be easily penetrated, while a light sludge, such as metal hydroxide slurries, will be soft and not resist

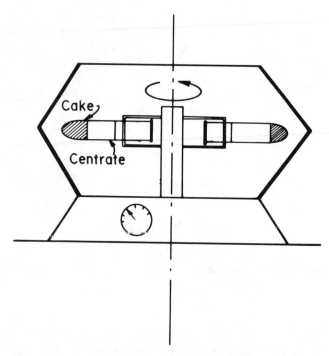

Figure 3.10.   A strobe light placed above a spinning test tube can be used to observe the settling of a sludge under gravitational force.

penetration, and will be correspondingly difficult to move out of a solid bowl centrifuge.[9]

Operation

The variables in the design and operation of a solid bowl centrifuge are listed in Table 3.2. The first group is machine variables, only some of which can be controlled by the operator. The most important variables at his command are in the second group, the process variables, and include flow rate and feed characteristics (e.g., solids concentration, chemical conditioning, age of sludge). He may also be able to change some of the machine variables, such as pool depth and bowl speed, but this is seldom done.

The centrifuge is to achieve two objectives: (1) produce a dry sludge cake, and (2) discharge a clear centrate. Unfortunately, the two objectives are difficult to attain simultaneously. In fact, the minor adjustments in any of the variables (with the exception of sludge characteristics) produce data which, when plotted on a graph of solids recovery vs cake solids concentration, will produce one line (Figure 3.12). For example, increasing the pool

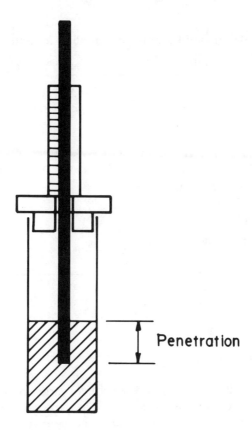

Figure 3.11. Penetrometer used for estimating the firmness of a sludge.

depth (thus decreasing the dry beach and increasing the hydraulic residence time) increases the solids capture. That is, since the residence time is greater, a larger fraction of the solids can settle out. This is, however, at the expense of cake solids, since the solids have less time to dewater on the beach, and since the wetter solids particles are captured and the water they carry contributes to the wetter cake.

Only by changing the characteristics of the sludge can both recovery and cake solids be increased (shift the line in Figure 3.12). The most common means of doing this is to condition the sludge chemically, usually with organic polyelectrolytes. Another way of changing the sludge characteristics is to adjust the sludge processes preceding the centrifuge. For example, a raw primary sludge in wastewater treatment dewaters fairly well with centrifugation, but will not dewater as well if it has gone septic. The age and condition of the sludge thus can have a significant effect on performance.

Table 3.2.   Design and Operational Variables for Solid Bowl Centrifuges

Machine Variables	Process Variables
Pool depth	Feed rate
Bowl length	Feed characteristics
Beach angle	solids concentration
Bowl speed	chemical conditioning
Conveyor speed	age of sludge
Number of leads and pitch of conveyor	
Bowl radius	
Point of chemical addition	
Pool depth	

Performance

The performance of solid bowl centrifuges varies considerably, since they can be operated even as thickeners. The best cake solids that can be expected, however, are for raw primary sludges at about 30% solids. Waste activated sludges rarely attain greater than 10% cake.

Filtration

Process

The filtration of sludges involves the use of a porous medium, such as a fabric, through which the liquid can pass, but the solids cannot. In some cases, such as with a gravity filter, no additional force is employed. Most filters, however, use either vacuum or positive pressure.

Hardware

The most common vacuum filter historically is the rotary drum filter (Figure 3.13). Modern rotary vacuum filters pick up the sludge out of a bottom trough, dewater it through a fabric (or metal coil) covering the perforated vacuum drum, and drop it off as the fabric is lifted off the drum and forced to go over a small roller.

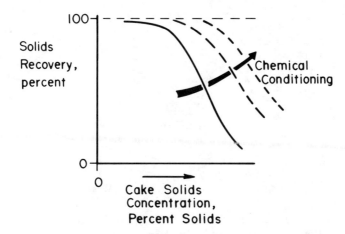

Figure 3.12.   Centrifuge performance tradeoff between cake solids concentration and solids recovery. Only by changing the sludge characteristics can the curve be shifted so as to increase both recovery and cake solids.

The standard vacuum filter does not perform well with light sludges, such as activated sludge and some metal hydroxides, even after chemical conditioning. The sludge will not be picked up by the filter cloth, and the pores of the fabric are plugged too rapidly (called "blinding"). Using a precoating technique allows for the performance to be improved, but this represents a significant operating expense.

The belt filter, first developed in Europe and introduced in the United States only 10 years ago, is a marked improvement over the rotary vacuum filter for dewatering difficult sludges. Although the various proprietary systems differ in detail, most belt filters employ the basic mechanisms illustrated in Figure 3.14. A chemically conditioned sludge first drops onto a perforated belt, where gravity drainage takes place. The already thickened sludge is next pressed between a series of rollers to produce a dry cake. A major problem with some types of sludge is the tendency to squirt out the sides of the belt as they are being squeezed.

In the filter press (Figure 3.15) a chemically conditioned sludge is pumped into cavities formed by a series of plates covered by a filter cloth. The liquid finds its way through the filter cloth, leaving the sludge solids behind. These finally fill up the cavity and become the sludge cake. The plates are opened up and the sludge is removed.

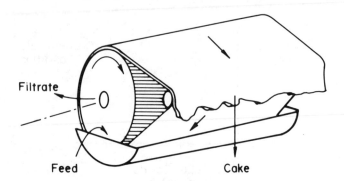

Figure 3.13. Rotary drum vacuum filter. A vacuum is drawn inside the drum at the sludge solids and deposited on top of the fabric covering.

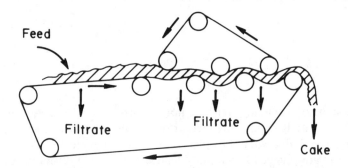

Figure 3.14. The belt filter employs gravity drainage and positive pressure.

In the past, the filter press did not find great popularity in wastewater treatment because it requires operator attention, especially during cake discharge. It does, however, exhibit one major advantage: given a sufficiently long filter run, it will dewater most sludges, and will produce a cake dryer than that attainable by any other mechanical process.

Design

Filters are sized on the basis of "filter yield," defined as the dry solids produced as cake per unit time per unit area of filter surface. Typically,

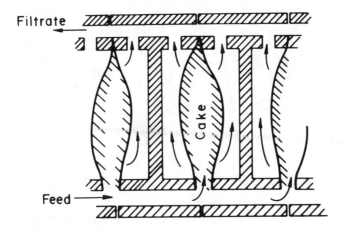

Figure 3.15.   The filter press has plates and frames which are covered by a filter cloth and through which the liquid flows as the sludge is pumped in under pressure.

filters used for dewatering municipal sludge achieve from 2 to 4 lb dry solids/hr/ft^2 of filter surface.

Filtration on a rotary vacuum filter involves several distinct steps: sludge pickup, sludge dewatering by filtration, and sludge cake discharge. These can be readily duplicated using a filter leaf apparatus (Figure 3.16). A grooved disc, covered with a filter fabric and connected to a vacuum source, is immersed into sludge for a specific time and then removed (sludge pickup) and held up in the air (dewatering). The vacuum is then released and the cake scraped off (sludge cake discharge). The cake solids are dried and weighed, and the amount of dry sludge determined. The "filter yield" is then calculated using the weight of solids and the area of the filter lead and the time of the total filtration cycle.

At the present time, no similar sizing techniques have become standard practice for belt or vacuum filtration. Some work has been done on both, however. For the belt filter, a simulation technique involving the pressing between two layers of fabric has been developed in England,[10] while pressure cells simulating pressure filters have been used by some researchers.[11]

In all filtration applications, it is advantageous to be able to characterize a sludge relative to its ability to be filtered, analogous to the "settleability coefficient" used for characterizing the ability of a sludge to dewater by centrifugation. The "specific resistance to filtration" concept has gained wide acceptance as a means of describing the filterability of sludge. As the specific resistance increases, the sludge becomes more difficult to dewater.

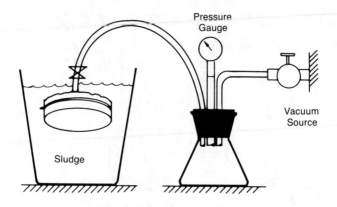

Figure 3.16.   Filter leaf apparatus used for sizing vacuum filters.

The apparatus shown in Figure 3.17 is used for measuring specific resistance to filtration. Sludge is poured into the Buchner funnel lined with a paper filter, and the rate of filtrate production measured in the graduated cylinder. The filtrate volume is plotted on arithmetic paper against the time divided by the volume (Figure 3.18). This (usually) results in a straight line with a slope defined as "b". The specific resistance is then calculated as

$$r = \frac{2\ PA^2b}{\mu w}$$

where P = vacuum pressure employed
A = area of the Buchner funnel
b = slope as defined above
μ = viscosity
w = cake deposited per volume of filtrate (usually estimated as the sludge solids concentration)

The measurement of specific resistance is somewhat cumbersome, especially if a series of tests is needed for estimating the effect of chemical conditioner. Such a series could run into hundreds of tests. A "quick-and-dirty" technique has been developed for estimating sludge filterability, based on the movement of water out of a sludge and into a blotter paper. As illustrated in Figure 3.19, this device measures the time necessary for the water in the blotter to move one centimeter, and the time can range from a few seconds for extremely well conditioned sludge, to several minutes for sludges which do not release their water readily. The "capillary suction time," or CST, is a useful tool for rapidly estimating the ability of a sludge

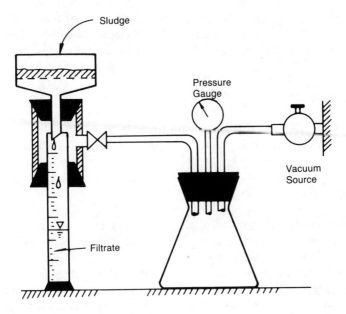

Figure 3.17. Buchner funnel apparatus for determining the specific resistance to filtration.

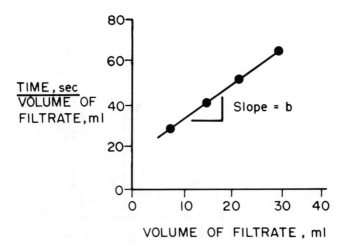

Figure 3.18. Typical results of a Buchner funnel test for measuring specific resistance to filtration.

to be rid of its water. It is *not*, however, a measure of how well a specific sludge would filter with a given filtration system, since other factors, such as sludge pickup, clogging of the filter cloth, and discharge, play equally important roles.

Operation

All filters require chemical conditioning. Because corrosion and abrasion are not serious problems, lime and ferric chloride are often used as the chemicals of choice.

Performance

Care must be taken in calculating the effectiveness of a filter if lime and ferric chloride are used, because they will contribute to the weight of cake solids. While 40% cake solids are possible with filter presses, often 10% of that 40% represents the $Ca(OH)_2$ added to the sludge. Especially where sludges are to be incinerated, a dry cake of 40% solids may not be enough to attain autogenous combustion if 10% of that represents inorganic lime.

Drying

The only reasonable means of drying is the classical "sand drying bed," which uses solar energy as its heat source.

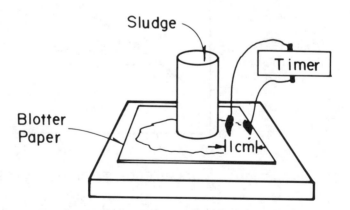

Figure 3.19. Capillary suction time apparatus.

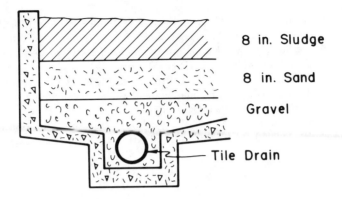

Figure 3.20. Cross section of a standard sand drying bed.

A typical sand bed is shown in Figure 3.20. The sand covers a tile field and the sludge is spread on the sand, usually about 8 in. deep. During the first dew days, the primary means of dewatering is filtration through the sand. As the pores are clogged up, however, the major liquid removal becomes evaporation. Typically, sludges lose enough liquid to crack, and this promotes further drying. In 30 to 60 days, municipal sludges are typically at solids concentrations of 30 to 40%. Covered beds can yield even drier solids if necessary.

REFERENCES

1. Vesilind, P. A. *Treatment and Disposal of Wastewater Sludges*, 2nd ed. (Ann Arbor, MI: Ann Arbor Science Publishers, Inc., 1979).
2. Jordan, V J., and C. H. Scherer. "Gravity Thickening Techniques at a Water Reclamation Plant," *J. Water Poll. Control Fed.* 42 (1970).
3. Dick, R. I., and B. Ewing. "Evaluation of Activated Sludge Thickening Theories," *J. San. Eng. Div.*, ASCE 93(SA4) (1967).
4. Vesilind, P. A. Discussion of Dick and Ewing, *J. San. Eng. Div.*, ASCE, 95(SA1) (1969).
5. Wood, R. F., and R. I. Dick. "Factors Influencing Batch Flotation Tests," *J. Water Poll. Control Fed.* 45(2) (1973).
6. Ambler, C. M. "The Evaluation of Centrifuge Performance," *Chem. Eng. Prog.* 48(3) (1952).
7. Vesilind, P. A. "Scale-up of Solid Bowl Centrifuge Perforance," *J. Env. Eng. Div.* ASCE 100(EE2) (1974).

8. Vesilind, P. A. "Characterization of Sludge for Centrifugal Dewatering," *Filt. & Sep.* (Br) March/April (1977).
9. Vesilind, P. A. "Estimation of Sludge Centrifuge Performance," *J. San. Eng. Div.* ASCE 96(SA3) (1970).
10. Barkerville, R. C., A. M. Bruce, and M. C. Day. "Laboratory Technique for Predicting and Evaluating the Performance of a Filterbelt Press," *Filt. & Sep.* (Br) September/October (1978).
11. Gerlich, J. W., and M. D. Rockwell. "Pressure Filtration of Wastewater Sludge with Ash Filter Aid," EPA R2-73-231, Washington, DC, 1974.

CHAPTER 4

SLUDGE DISPOSAL

INTRODUCTION

The two often conflicting concerns in sludge disposal are economics and environmental impact. Each of these factors also includes related constraints. For example, the ease of sludge disposal and uninterrupted operation are of importance to all operators. Thus reliability is a primary concern, and must be factored into the selection of disposal alternatives. The incorporation of these concerns is discussed in detail in the next section. In this section, disposal alternatives are discussed from the standpoint of technological availability and performance.

There are numerous alternatives to disposal, including the following:

- landfilling
- landspreading on agricultural land
- landspreading on reclaimed land
- land farming
- ocean disposal
- deep well injection
- contract hauling (or give-away)

These are discussed below.

LANDFILLING

Sludges containing hazardous materials at sufficiently high concentrations, such as sludges from several industries, must be disposed of in secure landfills. Typically, secure landfills must have either clay or synthetic rubberized liners, (PVC, polyethylene, etc.) to prevent seepage into the groundwater. Monitoring wells are often required to periodically test the ground-water around the landfill for potential contamination. In some landfills, the leachate is collected in a tile drainage system and treated.

Fortunately, municipal sludges are not classified as hazardous, and do not need to be placed into secure landfills. Most states do require, however, that the sludge be stabilized and dewatered prior to placement into a munic-ipal landfill or a dedicated landfill. The future of this operation is in doubt, as more and more states are moving toward abolishing all types of landfills.

LANDSPREADING ON AGRICULTURAL LAND

Sludge can be placed on agricultural land or other land where it can come into contact with the public and where health concerns are important. In the case of agricultural land, sludge represents a source of nutrients and acts as a soil conditioner. In addition, sludge may be placed on land which is either forested or has been badly impoverished, such as strip mines. Finally, sludge can be placed on dedicated land, where public contact is eliminated and the land is used for the sole purpose of assimilating the organics in the sludge into the soil.

Present thinking is that sludge should be applied to agricultural land at a rate equal to the nitrogen uptake of the crop in order to prevent nitrate contamination of groundwater. In most cases, this will govern the application rate. In other situations, however, the loading of metals will limit the amount of sludge applied. Of particular concern is cadmium.

Nutrients

Sludges contain appreciable amounts of nitrogen (N) and phosphorus (P), but very little potassium (K), which is leached from the sludge and released in the plant effluent. Table 4.1 is a summary of the concentrations of N and P in sludge. It is interesting that sludge nutrient values are very similar to those of manure.

Nitrogen may be in any one of four forms: organic-N, NH_3-N, NO_2-N, and NO_3-N. The latter three are available to the plants as nutrients, while soil microorganisms must convert the organic-N first to one of the inor-ganic forms before it becomes available. Eventually about 80% of all nitro-gen in sludge becomes available for the plants.

Phosphorus is available at about the same rate as in inorganic fertilizers, and varies according to the soil and plants. Typically the relative P utiliza-tion rate is about 40 to 80%.

There is some question as to which nutrient — nitrogen or phosphorus — should be the limiting nutrient in sludge application. Although most stan-

Table 4.1. Nitrogen and Phosphorus in Sludge and Comparable Organic Wastes

Type of Organic Waste	N (kg t^{-1} DM)	P (kg t^{-1} DM)
Farm yard manure DK[a]	30	10
Farm yard slurry	75	10
Human excreta	250	36
Urban sludge, mixed	~40	~20
Urban sludge, U.S. range[b]	1–176	1–143
Urban sludge, U.S. median[b]	33	23
Urban sludge, U.S. mean[b]	39	25

[a]From reference 1.
[b]From reference 2.

dards in the United States rely on the nitrogen loading (because of the fear of groundwater contamination with nitrate), Hansen et al.[3] have argued that it is more reasonable to establish the loading rate on the basis of phosphorus.

Cadmium

Cadmium is a toxic metal which has no known benefit in human metabolism, and which can be translocated from sludge to soil to plant to animal to human. Table 4.2 shows some typical concentrations of cadmium in sludge. Obviously the concentration of cadmium varies with the type of wastewater discharged. It is possible to calculate, based on Cd concentrations in human waste, that if the only source of cadmium were in sanitary sewage, the concentration of cadmium should be about 0.3 ppm in sludge. This figure should serve as a goal for pretreatment programs, since it would seem unlikely that Cd concentrations in sludge could be reduced below this level.

Hansen et al.[4] have shown, based on both empirical and theoretical evidence, that plant uptake of Cd is linearly correlated to total soil Cd. Their results are shown in Figure 4.1. It is noted that pH is a dominating parameter as to plant uptake of Cd. Guidelines as to maximum permissible Cd addition would be different depending on the soil. Thus it seems that limits can be translated in terms of the Cd in the soil, and not only on the rate of Cd application to the soil.

Table 4.2.   Cadmium in Sludge (in ppm[a])

Range	Median	Mean
3–3410	16	110

[a]1 mg Cd per kg dry solids = 1 ppm.
Source: Reference 2.

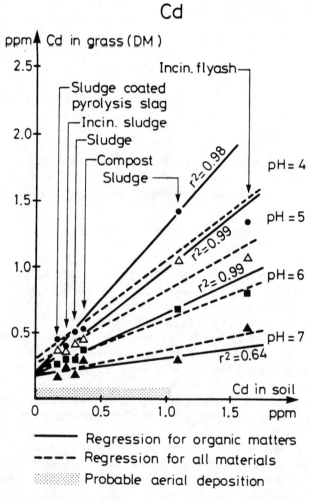

Figure 4.1.   Cadmium balance in solids.[4]

Pathogens

Pathogens in sludge imply the obvious risk of spreading infectious disease when trying to recover nutrients from sludge in agricultural sludge utilization. There is evidence that such disease transmission will occur when sludge is not properly managed.[5]

The three types of microorganisms that can be troublesome are bacteria, parasites, and viruses.

Salmonellae is the predominant group of bacteria in sludge, and numerous species are present which can cause human disease. Because of this, *salmonellae* have often been suggested as the appropriate microorganism to act as an indicator of sludge disinfection. *Salmonellae* can be inactivated at the treatment plant by any number of techniques (see Section 2), including composting and pasteurization. Such processes as aerobic and anaerobic digestion do not eliminate the risk of *salmonellae* infection, but reduce it markedly. Once on the soil, pathogenic bacteria are inactivated, and this rate is dependent on soil temperatures. Typically, 20 weeks is the limit of pathogen survival.

The parasites in sludge include the ova of the beef tapeworm and various nematodes and ascaris species. Because a single embryo is sufficient to cause infection, inactivation before ingestion is necessary. As with bacteria, digestion will reduce the numbers of parasites, but will not destroy them. Their survival in the soil is longer than for bacteria. Ascaris eggs have been known to survive as long as 14 months.

Many different pathogenic viruses are known to exist in sludge, including polio and hepatitis. Little information is available on their transmission, and the inactivation in a treatment plant generally follows the pattern of bacterial inactivation.

LANDSPREADING ON RECLAIMED LAND

The amount of sludge applied to reclaimed land is calculated on the basis of nitrogen loading. There are no limits on the metals or other contaminants.

LAND FARMING

Land farming is the placement of sludge on a dedicated plot of ground for the sole purpose of using the soil microorganisms to assimilate the sludge solids. Since the land is not generally of high economic value, the sludge can be placed without undue potential for health problems or environmental damage. Often sludge is sprayed onto the land, since dewatering is an unnecessary expense. When dewatering is practiced, placement of sludge into landfills is generally more economical.

Where the sludge could produce odor problems, or where the assimilative capacity of the ground surface is too low, subsurface disposal is practiced. Tank trucks with subsurface injection systems are used to place sludge about 10 to 20 cm below the surface, where aerobic microorganisms easily assimilate the organics into the soil structure.

The various methods for applying liquid sludge on the surface of land, and for subsurface disposal, are shown in Table 4.3. The disposal of dewatered sludges is tabulated in Table 4.4.

Table 4.3.   Surface and Subsurface Application Methods and Equipment for Liquid Sludges[a]

Method	Characteristics	Topographical and Seasonal Suitability
Surface Applications		
Spray (sprinkler) fixed or portable	Large orifice required on nozzles; large power and low labor requirement; wide selection of commercial equipment available; sludge must be flushed from pipes when use stops for longer than 2 to 3 days.	Can be used on a sloping land; can be used yearround if the pipes are drained in winter; odor and aerosol nuisances may occur.
Overland flow and flooding	Used on sloping ground, with or without vegetation, with no runoff permitted; suitable for emergency operation; difficult to get uniform aerial application; use of gated or perforated pipe requires screening of sludge prior to application; sludge must be flushed from	Can be applied from all-weather ridge roads.

Table 4.3., continued

Method	Characteristics	Topographical and Seasonal Suitability
	pipes when use stops for longer than 2 to 3 days.	
Ridge and furrows	Land preparation needed; lower power requirements than spray; limited to low solids concentration (less than 3% works best).	Between 0.3 and 1.0% slope depending on solids concentration and condition of soil. Fillable land not usable on wet or frozen ground.
Tank truck	Capacity 500 to 3,800 gal; larger volume trucks will require flotation tires; can use with temporary irrigation setup; with pump discharge can spray from roadway onto field.	Tillable land; not usable on very soft ground.

Subsurface Applications

Method	Characteristics	Topographical and Seasonal Suitability
Flexible irrigation hose (umbilical cord system) with subsurface injection or surface discharge[b]	Pipeline or tanker pressurized supply; 650-ft hose connected to manifold discharge on plow or disk pulled by tracked vehicle; abrasive wear can result in short hose life; subsurface injection by means of very small furrow behind knife-edge cutting disk and/or narrow plow; surface discharge into furrow immediately ahead of plow; application rate of 50 to 100 wet ton/acre/pass.	Tillable land; not usable on wet or frozen ground.
Tank truck with subsurface injection or	500- to 3,800-gal 4-wheel drive commercial equipment available;	Tillable land; not usable on wet or frozen ground.

Table 4.3., continued

Method	Characteristics	Topographical and Seasonal Suitability
surface discharge[b]	subsurface injection by means of very small furrow behind knife-edge cutting disk and/or narrow plow; surface discharge into furrow immediately ahead of plow; application rate of 50 to 100 wet ton/acre/pass.	
Farm tank trailer and tractor with surface discharge[b]	Sludge discharged into furrow ahead of plow mounted on tank trailer; application of 170 to 225 wet ton/acre/pass. Sludge spread in narrow bank on ground surface and immediately plowed under; application rate of 50 to 125 wet ton/acre/pass.	Tillable land; not usable on wet or frozen ground.
Farm tank trailer and tractor with subsurface injection[b]	Sludge discharged into channel opened and covered by a tillable tool mounted on tank trailer; application rate 25 to 50 wet ton/acre/pass.	Tillable land; not usable on wet or frozen ground.

[a]1 gal = 3.8 liters; 1 ton/acre = 2.25 t/ha.
[b]Vehicle reaccess to area receiving application dependent on water content and application rate of liquid sludges.
Source: Reference 5.

OCEAN DISPOSAL

For those municipalities close enough to a large body of water, such as a gulf or ocean, the disposal of sludges into these waters is an alternative. At the present time, however, the EPA strongly discourages ocean disposal,

Table 4.4.   Methods and Equipment for Application of Dewatered Sludges

Method	Characteristics
Spreading	Truck-mounted or tractor-powered box spreader (commercially available); sludge spread evenly on ground; application rate controlled by over-the-ground speed; can be incorporated by disking or plowing.
Piles or windrows	Normally hauled by dump truck; spreading and leveling by bulldozer or grader needed to give uniform application; 4 to 6 inch layer can be incorporated by plowing.
Reslurry and handle as in Table 4.3.	Suitable for long hauls by rail transportation.

Source: Reference 5.

and only a few cities, by means of court injuctions, are still continuing the practice. It is unlikely that ocean disposal will be a reasonable alternative in the foreseeable future, and thus it is not discussed further in this book.

DEEP WELL INJECTION

Sludge disposal has in the past been accomplished by injecting the slurry into deep wells and storing it below ground. This technique is especially attractive in cases where deep wells are already available (such as oil fields). It also has the advantage of being a final disposal step, in that it is unlikely that the waste would come into contact with humans in the future.

Unfortunately, this has not always been the case. The injection of toxic and otherwise hazardous chemicals into lower geological strata involves considerable risk. Probably the most insidious aspect of deep well injection is that it is for the most part irreversible. If it is found that a well is indeed contaminating an aquifer or reemerging at the surface some distance away from the well, there is little that can be done to correct the situation. Once the waste material is in the ground, it is unlikely that it can ever be extracted. For that reason, the possible adverse environmental impact of deep well injection is quite serious.

CONTRACT DISPOSAL

Contract disposal — the purchase of disposal services — is not a disposal technique as such, since it simply transfers the problem to another location and person. From a practical vantage point, however, it is often the least expensive way. Because of stricter federal and state regulations, disposal services are becoming increasingly available, and economically competitive with alternative schemes. For small communities, such disposal services often represent the best combination of economics, dependability and convenience.

REFERENCES

1. Hansen, J. A., and J. Tjell. "Agricultural Use of Sludge and Compost," Environmental Contamination in Context. EWPCA-ISWA Symposium '84, Munich 21–25 May, 1984.
2. "Land Application of Sludge Process Design," Manual, EPA-625/1-83-016 MERL, U.S. EPA Cincinnati, OH, 1983.
3. Hansen, J. A., J. C. Tjell, and R. Lassen, "Sanitary Phosphorus Recovery for Agriculture," International Conference on Harmless Disposal of Communal and Other Organic Wastes with Special Regard to Farm Land Application. Hungarian Hydrological Society, Budapest, Hungary, May 30–June 4, 1983.
4. Hansen, J. A., J. C. Tjell, and T. Christensen. "The Risk to Health of Microbes in Sewage Sludge Applied to Land," Report of a WHO Working Group, Stevenage 6-9 January, 1981. WHO Regional Office for Europe, Copenhagen, Denmark.
5. "Process Design Manual for Sludge Treatment and Disposal," EPA 625/1-79-011, U.S. EPA Cincinnati, OH, 1979.

CHAPTER 5

EVALUATION OF ALTERNATIVES

INTRODUCTION

It is difficult to imagine another social problem which carries with it a more complex plethora of social, environmental and economic concerns than the disposal of municipal sludges. The engineer must successfully negotiate through this morass, and develop a plan which he/she can justify as meeting the highest number of positive criteria, including of course the least cost for the client. This analysis is difficult and open to debate, since such criteria as long-range degradation of farmland, destruction of marine ecosystems, or transmission of human disease are factors which cannot readily be quantified in terms of dollars. Thus it is necessary to approach the sludge disposal problem in a systematic and orderly manner, making small decisions which can be supported, and developing and ranking alternatives which will stand the test of social scrutiny. One of these techniques is the idea of transformation curves, discussed below.

TRANSFORMATION CURVES*

Consider two objectives, both of which have value, and a number of alternatives, all of which may fulfill these objectives. How can one alternative be selected from among the possibilities in such a way as to assure maximization of the objectives?

---

*This technique for the evaluation of multiple variables is based on "Analytical Aids to Decision Making: Measures of Effectiveness and Multiple Objective Trade-off Displays," by Myron S. Rosenberg, Ph.D., P.E., Massachusetts Institute of Technology, 1980."

Beginning with a trivial example, suppose the objective is to purchase
bubble gum such that the bubbles will be the biggest possible. There are a
number of different bubble gum types on the market, and they vary in
price.

One variable is clearly the size of bubble produced with a given gum, with
the smallest diameter being worst. The other variable is cost, but increased
cost is not better. As the cost numbers increase, the advantages to the
purchaser decrease. This problem can be resolved by transforming the cost
numbers by taking their reciprocal, so that, numerically, high values of 1/
cost are better than low values of 1/cost.

The cost, 1/cost, and bubble diameter, can now be tabulated for each
brand of bubble gum, and the values of diameter and 1/cost can be plotted
as shown in Figure 5.1. The curve connecting the outermost points, which
must always have a negative slope, describes the transformation curve.
One's objectives are maximized by all alternatives which appear on the
curve. In the bubble gum example, gums number 3 and 5 meet this criterion,
and thus would be the two gums of choice.

But suppose it would not be possible to purchase gum which cost more
than 25 cents regardless of how big the bubble was. In other words, the
possible transactions are limited by the availability of ready cash. This can
be reflected in the analysis by a "bound," or a straight line drawn at 1/25, so
that all data points below this line are not in the feasible range of options.

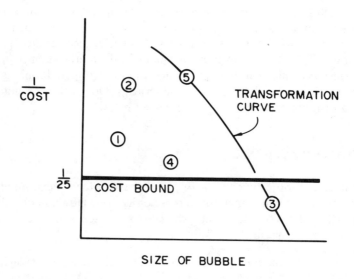

Figure 5.1. Typical transformation curve.

Thus gum number 3, even though it can produce the greatest bubble, is not a feasible solution to the problem.

Often there are more than two objectives. It is necessary to maximize utility, for example, minimize operating cost, and maximize public acceptance. All of these are legitimate variables, and cannot be analyzed on a two-variable coordinate plot. The transformation curve analysis incorporates multiple objectives by drawing curves for each pair of objectives and noting which alternatives are placed on the transformation curve for each plot.

For example, the data in Figure 5.2 show that four objectives are to be studied: capital cost; operating and maintenance cost; comparative reliability and implementability; and non-cost factors which might incorporate aesthetics, environmental concerns, etc.

The first curve, where the reciprocals of capital costs are plotted against the reciprocals of operating costs, shows that alternatives 1,3,4,6, and 7 are on the transformation curve, but alternative 7 is below the capital cost bound and thus not feasible.

Next, the costs are combined and expressed as present worth, and the reciprocal of present worth plotted against the reliability/implementability, and against the index of non-cost factors. From these two plots, alternatives 4 and 5, and 4,5, and 6, respectively, appear to be feasible.

Combining the results of these curves, as in Table 5.1, it is noted that only alternative 4 is represented in all transformation curves, and thus becomes the alternative of choice.

In sludge disposal, the engineer must develop an index for the reliability/implementability of alternative systems, and also construct a non-cost

Table 5.1. Summary of Transformation Curve Analysis Shown in Figure 5.2.

	Symbol on the Transformation Curve		
Alternative	Capital Cost/ O&M Cost	Present Worth/ Reliability	Present Worth/ Non-Cost Factors
1	X	–	–
2	–	–	–
3	X	–	–
4	X	X	X
5	–	X	X
6	X	X	–
7	–	–	–

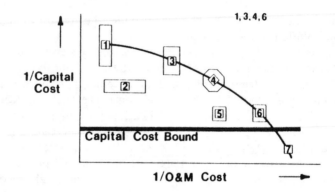

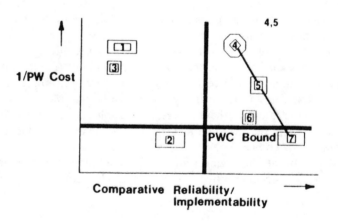

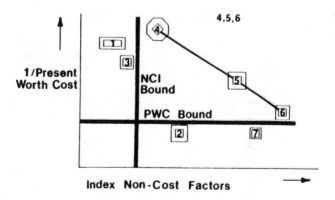

Figure 5.2. Transformation curves with multiple objectives.

index. This can be done most effectively with the client by asking him/her to rate the relative importance of various factors, such as time spent on maintenance, odor potentials, etc. Thus the client feels that the concerns of the community have been fully expressed in the analysis. In Part II of this book, an example of transformation curves applied to a real sludge disposal problem is presented, but it is not suggested that the actual rating schemes used are the most applicable to all communities. The need for communicative and engineering skills in designing the rating scheme and questions will become quite evident when the problem is formulated.

CHAPTER 6

REGULATORY CONSTRAINTS

INTRODUCTION

Without regulatory constraints, sludge disposal would involve the dumping of sludges in places and times which would be the most expedient to the wastewater treatment plant operators and which they could "get away with." The classic example of this is a wastewater treatment plant operator whose plant along Lake Ontario had periodic "accidents," like someone leaving the valve from the digester open overnight. With regulatory constraints, however, the options for sludge disposal are strictly limited.

APPLICABLE FEDERAL LAWS*

Clean Water Act

CWA (Clean Water Act of 1977, PL 95–217 and the Federal Water Pollution Control Act of 1972, PL 92–500) authorizes federal funding of 75% (85% for innovative and alternative technology projects) of the eligible costs involved in the construction of municipal wastewater treatment plants and sludge treatment and disposition facilities; authorizes EPA to issue comprehensive sewage sludge management guidelines and regulations; authorizes the NPDES (National Pollution Discharge Elimination System) for point source discharges and development of areawide waste treatment

---

*This discussion is adapted from an EPA document: *A Guide to Regulations and Guidance for the Utilization of Disposal of Municipal Sludge*, EPA 430/9–80–015, 1980.

or water quality management plans for non-point source pollution; requires the implementation of pretreatment standards for industrial discharges that enter POTWs; and establishes a major research and demonstration program to develop improved wastewater treatment and sludge management practices.

Resource Conservation and Recovery Act

RCRA (Resource Conservation and Recovery Act of 1976, PL 94–580) provides financial assistance to state and local governments for development of solid waste management plans which provide for the safe disposal of solid waste; provides that technical assistance be provided to help establish acceptable solid waste management methods; requires regulations for the safe disposal of hazardous and nonhazardous wastes; and encourages the research and demonstration of more effective solid waste disposal and resource conservation technologies.

Marine Protection Research and Sanctuaries Act

MPRSA (Marine Protection Research and Sanctuaries Act of 1977, PL 92–532) effectively phases out ocean disposal of sewage sludge by December 31, 1981. MPRSA also gives EPA the authority to determine a reasonable compliance schedule for the implementation of landbased disposal alternatives.

Clean Air Act Amendments

CAA (Clean Air Act Amendments of 1970 and 1977, PL 91–604 and PL 95–95) authorized the development of State Implementation Plans (SIPs) for the purpose of meeting federal ambient air quality standards. To meet the CAA objectives, EPA has developed an emission offset policy for new or modified incinerator and heat drying facilities and a procedure for preventing the significant deterioration of ambient air quality. CAA also authorizes regulations for the control of hazardous air pollutants and new source performance standards.

Safe Drinking Water Act

SDWA (Safe Drinking Water Act of 1975, PL 93–523) requires coordination with the CWA and RCRA to protect drinking water from contamination.

National Environmental Policy Act

NEPA (National Environmental Policy Act of 1969, PL 91–190) authorizes Regional Administrators, at their discretion, to require Environmental Impact Statements (EIS) (40 CFR, Part 6) if potential adverse social, economic or environmental impacts are suspected for a new or modified sludge disposition facility or practice. An EIS or negative declaration (40 CFR, Part 35, Sect. 35.925–8) is also required when applying for Federal Construction Grants.

Toxic Substances Control Act

TSCA (Toxic Substances Control Act of 1976, PL 94–469), Section 9, requires coordination with the Clean Air Act and the Clean Water Act to restrict disposal of hazardous wastes. Presently only PCB (polychlorinated biphenyl) is specifically regulated in regard to sludge disposition.

APPLICABLE FEDERAL REGULATIONS

The following regulations apply to the various sludge disposition methods.

Criteria for the Classification of Solid Waste Disposal Facilities and Practices

Criteria for the Classification of Solid Waste Disposal Facilities and Practices (40 CFR, Part 257; *Federal Register*, Sept. 13, 1979) is authorized by Section 405(d) of the CWA and 4004(a) and 1008(a)(3) of the RCRA. Guidance Manual SW–828 should be consulted to determine if a facility complies with the Criteria.

National Pollution Discharge Elimination System

NPDES (National Pollution Discharge Elimination System, 40 CFR, Part 125) is authorized by Section 402 of the FWPCA.

Federal Construction Grants Regulations

Federal Construction Grants Regulations (40 CFR, Part 35, Subpart E; *Federal Register*, Sept. 27, 1978) are authorized by section 201 of the CWA.

State or Areawide Waste Treatment Management Plans

State or Areawide Waste Treatment Management Plans are authorized by Sect. 208 of the CWA.

Air Regulations

Air Regulations are authorized by the CAA.

Ocean Dumping Regulations

Ocean Dumping Regulations are authorized by MPRSA and CWA.

Hazardous Waste Regulations

Hazardous Waste Regulations (40 CFR, Parts 260–265; *Federal Register*, May 19, 1980).

HAZARDOUS WASTE REGULATIONS

Subtitle C of RCRA authorized the development of hazardous waste regulations. Under the proposed hazardous waste regulations, issued on December 18, 1978 in the *Federal Register*, municipal sewage sludges were excluded from coverage under Subtitle C of RCRA. Subsequently, in the final regulations promulgated in the *Federal Register* on May 19, 1980, municipal sewage sludges were no longer excluded from coverage, and thus are potentially subject to control as hazardous waste.

Domestic sewage and any mixture of domestic sewage and other wastes that passes through a sewer system to a POTW for treatment is not considered a solid waste [40 CFR Part 261.4(a)(1)]. Under all circumstances, however, municipal sewage sludge that is separated from the sewage during treatment is considered a solid waste [261.2(a)]. In general, a solid waste is a hazardous waste if it has been listed as such by the Administrator or if it exhibits any of the defined characteristics of a hazardous waste [261.3(a)].

EPA has not listed municipal sewage sludges as hazardous wastes. Therefore, municipal sewage sludges are not considered hazardous unless tested and shown to be hazardous. While not included in the Agency's listing of hazardous wastes under Subpart D, of Part 261, specific municipal sewage sludges will be considered hazardous if they exhibit any one of the four characteristics of hazardous waste [261.21 through 261.24, i.e., ignitability, corrosivity, reactivity, and EP (extraction procedure) toxicity]. Specific municipal sewage sludges would also be considered hazardous if they were mixed with any hazardous waste other than those entering the publicly owned treatment works (POTWs) through a sanitary sewer system [261.3(a)(2)(ii) and 261.4(a)(1)(ii)]. To date, very few municipal sludges have been classified as hazardous waste.

Municipalities have an obligation to determine if their sludge meets the definition of a hazardous waste. This does not mean that each POTW must test its sludge. Rather, POTWs or other waste handlers must make a determination that the waste is not hazardous, based upon knowledge of the waste, including the contaminants, etc. EPA advises testing, particularly EP toxicity testing, where there are significant contributions of industrial wastewater or stormwater into the POTW, or where there is any reason to believe that the sludge may exhibit the EP toxicity characteristic. EPA believes that POTW sludge will rarely, if ever, exhibit the other three characteristics of a hazardous waste and that a determination can be made based on knowledge about the sludge, without need of testing.

The regulations place the responsibility of determining whether a POTW sludge is a hazardous waste squarely on the owner or operator of the POTW. He may choose any method he likes to make this determination. If he determines that his sludge is not a hazardous waste, or fails to make a determination, and EPA finds that the sludge is a hazardous waste, then he is in violation of the regulations.

The characteristic most likely to cause a sludge to be hazardous would be toxicity, determined by the extraction procedure. In very limited tests, cadmium is the only known element that has caused a sludge to fail the EP, i.e., be considered hazardous.

Any POTW that generates or transports a municipal sewage sludge which it believes to be hazardous and that plans to continue to generate, transport, treat or dispose of more than 1000 kg at any time, must notify EPA of its activity. A POTW which is only a generator of a hazardous municipal sewage sludge and does not also treat, store, or dispose of the sludge does not require a hazardous waste permit. This POTW generator, however, does have a major responsibility to follow all the provisions of 40 CFR Part 262. A POTW would also require a hazardous waste permit if it engaged in treatment, storage, or disposal of hazardous municipal sludge in the quantities described above. As part of this permitting process, an existing POTW must obtain interim status as a hazardous waste treater, storer, or disposer. To obtain this interim status the applicant POTW would have had to notify EPA and submit a completed Part A permit application to the appropriate EPA regional office.

If interim status has not been obtained, then the POTW would not be able to operate under the interim status provisions of the hazardous waste regulations. The POTW would have to notify EPA that it is generating a hazardous sludge. The POTW would also have to comply with the applicable requirements of Parts 262 and 263. Finally, if the POTW treats, stores, or disposes of its hazardous sludge onsite, then it must submit Part A and Part B of a permit application in accordance with Section 122.26(b). And, because the POTW does not have interim status, it would have to refrain from treating, storing, or disposing of its sludge onsite until it was issued a RCRA Subtitle C permit. While waiting for the issuance of a permit, the POTW would have to send its hazardous sludge to a hazardous waste treatment, storage, or disposal facility that has interim status or has been issued a RCRA Subtitle C permit.

The lack of interim status may present a very difficult problem for POTWs caught in this predicament. This is because (a) it will take time to issue a permit, (b) in the interim, it forecloses onsite digestion, dewatering and storage (except 90-day accumulation) of the large volumes of sludge typically generated by a POTW and (c) it requires unanticipated off-site transportation of the sludge to hazardous waste facilities that may not be available or may be located long distances away. EPA is currently examining the unique problems of POTWs regarding compliance with this provision. Any POTW that generates, treats, transports, stores, or disposes of a hazardous municipal sewage sludge without filing the notification is subject to civil or criminal penalties.

In the Appendix to this book a summary is presented of these regulations, along with some further interpretations. The Appendix also contains a summary of the State of Florida regulations relative to sludge disposal.

CHAPTER 7

CONCLUSIONS

INTRODUCTION

A wastewater treatment plant exists for the purpose of producing an essentially clear effluent which can be discharged to the environment without adverse impact. The design engineer and plant operator therefore are mostly concerned with the liquid processing within the plant, and solids management becomes important only when the plant malfunctions. Often, these upsets can be avoided by following a few fairly simply guidelines for sludge management within a plant. Malfunctions, from the operator's viewpoint, are either effluent quality problems which are forcefully brought to his attention by state or federal inspectors, or sludge problems which usually consist of having too much of a bad thing and nowhere to put it. The latter is often caused by poor sludge management practices, starting with the initial design of the plant. Following are some rules design engineers (and operators) should consider.

THREE LAWS OF SLUDGE MANAGEMENT

Stated concisely, these laws of sludge management are:

1. Don't hold sludges.
2. Don't mix sludges.
3. Don't recirculate sludges.

As is the case with any generality, these are not always applicable. They have been found to be true at enough plants, however, to warrant discussion.

HOLDING SLUDGE

Holding sludge (saving it, storing it, etc.) is usually the doing of a plant operator, and the procedure is often dictated by the vagaries of the labor force. It should nevertheless be the aim of the designer to allow the operator not to have or hold sludge before a subsequent operation, such as dewatering. Although this principle seems to be especially applicable to raw primary sludges, the holding of digested sludge (and the resulting cooling) has been shown to cut the capacity of mechanical dewatering to one third of the original.[1]

One clear outward sign of impending trouble is bubbling primary clarifiers, due to holding raw sludge in the primary clarifiers and the subsequent gas formation within the settled raw sludge. This will decrease the efficiency of solids removal, as well as place an extra load on the alkalinity of the anaerobic digesters. Sludge should be pumped from the primary clarifiers at sufficiently frequent intervals so as to avoid septicity.

MIXING SLUDGE

The second axiom — "Don't mix it" — is violated by both design engineers and plant operators. For example, experience seems to indicate that the discharge of excess waste activated sludge to the head of the plant results in deterioration of the primary clarifier performance, significantly increases the volume of sludge pumped, and necessitates more frequent pumping due to increased septicity problems.

It has been known for some time that when a raw primary sludge is mixed with waste activated sludge, the mixture assumes many of the undesirable characteristics of waste activated sludge. It is, for example, very difficult to mechanically dewater a 1/1 mixture of raw primary/waste activated sludge. The excellent dewaterability of raw primary sludges is sacrificed when mixed with secondary sludges.

It is, however, necessary to achieve the highest solids concentrations possible before a process such as anaerobic digestion, since the governing operational parameter in anaerobic digestion is solids retention time. One means of achieving this has been to blend the raw primary and waste activated sludges in a gravity thickener.[2] This has at times caused operational problems such as floating sludge and odors, mainly due to the oxygen-starved nature of the sludges. A more reasonable solution seems to be the gravitational thickening of primary sludges and the use of flotation thickeners for aerobic secondary sludges, with blending following the separate thickening operations and then subsequent anaerobic digestion. In plants where this type of scheme has been instituted, the solids dewatering has improved markedly, and operational problems have been reduced.[3]

Another example of improved operation by not mixing sludges is the scheme of using anaerobic digestion for raw primary sludge only, and employing aerobic digestion for the waste activated sludge. These stabilized sludges can then be blended and dewatered at a cost significantly lower than the alternative of mixing prior to stabilization.

RECIRCULATING SLUDGE

The final axiom — "Don't recirculate sludge" — is probably the most difficult of the three laws to promote, since the recirculation of various process streams to the head of the plant is almost a knee-jerk reaction for design engineers. These streams can, however, impose a significant excess solids load on the process.

In addition to the total magnitude of the additional solids load on the plant, the types of solids recirculated are almost exclusively the fine, difficult to handle solids. More than one treatment plant operator has found himself inundated with these fine solids, at the eventual expense in treatment efficiency. The only solution in such cases is to clean the plant out. In the design and operation of a plant, therefore, one should not recirculate solids if at all possible. Many operators prefer to dewater all of the sludge, and to do this at a high rate of solids recovery, recognizing that this investment is highly profitable for future plant operation.

CONCLUSION

Efficient and effective operation of wastewater treatment facilities requires careful consideration of sludge handling practices. It is suggested that three practical guidelines which could assist in the management of wastewater sludges are (1) don't hold sludges; (2) don't mix sludges; and (3) don't recirculate sludges.

REFERENCES

1. Vermehen, P. I. "Chemical Removal of Nutrient Salts from Plant Effluent," Sixth Nordic Symposium on Water Research, Copenhagen (1978).
2. Torpey, W. N. "Concentration of Combined Primary and Activated Sludges in Separate Thickening Tanks," Proc. ASCE 80, Sept., No. 443, (1954).
3. *Process Design Manual for Sludge Treatment and Disposal*, Technology Transfer, EPA 625/1–74–006, October (1974).

PART II

SLUDGE DISPOSAL FOR ST. PETERSBURG: A CASE STUDY

SECTION I
INTRODUCTION

1.1   PURPOSE AND SCOPE

The purpose of this report is to recommend to the City of St. Petersburg a long-term, cost-effective and implementable municipal sludge disposal program.

The City of St. Petersburg is located at the southern tip of Pinellas County as shown on the study area map, Figure 1-1. The City owns and operates four (4) wastewater treatment plants. The waste sludge from these plants must be disposed of, or utilized in an appropriate manner. In this report, various feasible alternatives for sludge disposal are reviewed, and the institutional arrangements and financial requirements are described and evaluated.

The suggested alternatives are limited to sludge treatment and disposal following sludge dewatering by belt filter presses already in place and operating at each of the four wastewater treatment plants (Figure 1-2). The degree and manner of sludge treatment recommended for each alternative reflects the requirements of the ultimate disposal proposed.

A regional sludge management solution is not within the scope of this study. Each alternative considers the disposal of sludge only from the City of St. Petersburg, except for the City/County Incinerator alternative.

1.2   ORGANIZATION

Following an assessment of the existing sludge treatment and production processes at the four wastewater treatment plants, and a review of the sludge characteristics as they relate to and restrict the ultimate disposal alternatives, various technologies were screened for appropriateness and suitability. Using the projected sludge quantities and qualities and suitable technologies, long-term alternatives were developed. These alternatives include City owned and operated facilities, full service contracts, a City/County incinerator alternative and the continued use of contract hauling. The alternatives judged to be most appropriate were then studied in detail as to cost-effectiveness, energy intensiveness, reliability, flexibility, technology and environmental impact.

In addition, the alternatives were judged on the basis of institutional requirements, availability of sites, market potential of the finished product, and related non-cost factors. Finally, a recommended alternative is proposed for implementation by the City of St. Petersburg.

This report is organized as follows:

Section 1   Introduction
Section 2   Background
Section 3   Sludge Quantities and Characteristics
Section 4   Available Technology
Section 5   Technology Assessment
Section 6   Alternative Approaches
Section 7   Development and Description of
            Alternatives

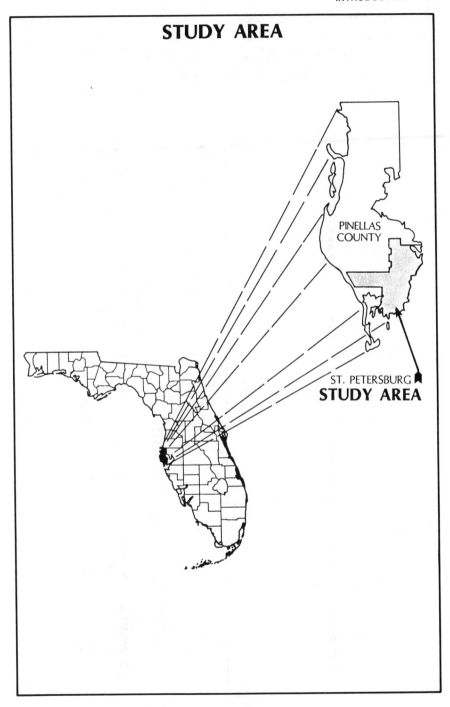

STUDY AREA

PINELLAS COUNTY

ST. PETERSBURG
STUDY AREA

Figure 1-1

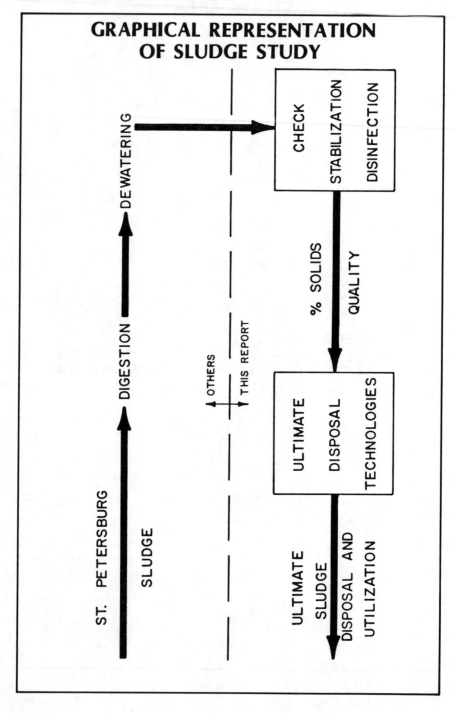

Figure 1-2

Section 8   Comparative Analysis
Section 9   User Costs and Payout Period Analyses
Section 10 Conclusions and Recommendations

1.3   METHODOLOGY FOR TECHNOLOGY ASSESSMENT

Manufacturers of each of the listed technologies were contacted in order to estimate the number of installations handling sludge and the current status of the technology. Preliminary estimates for capital costs were submitted by the manufacturers of each process based upon information and criteria supplied by the engineer. More detailed cost estimates and design data were developed for alternatives selected in Section 5 for further comparative analysis. This supplementary information included:

* operation cost
* maintenance cost
* storage requirements
* site requirements
* pollution control requirements

In order to evaluate the possibilities for full service Contract, the Full Service Operator submittals to the City of Orlando for sludge handling were reviewed.

SECTION 2

BACKGROUND

2.1   GENERAL BACKGROUND

The City of St. Petersburg operates four wastewater treatment plants which produce approximately 40,000 pounds (dry solids) of sludge each working day or 29,000 pounds (dry solids) per calendar day. This sludge is stabilized by anaerobic digestion and dewatered by belt press filters. For the City of St. Petersburg, the historical method of wastewater sludge disposal was hauling liquid sludge to either the Toytown Landfill or to a sludge lagoon near the City's northern limits. In 1983, the Toytown Landfill was closed to the liquid sludge and the sludge lagoon was reaching its capacity. With this impending problem, the City of St. Petersburg built sludge dewatering facilities and contracted out sludge hauling.

The ultimate disposal of this material presents a major problem and cost for the City. As shown in Figure 1-1, the City is bounded by water and is densely developed, thus making land disposal within City limits not feasible, and requiring a long haul distance to land disposal sites outside the City. Recognizing the dilemma presented by its location, the City has sought innovative approaches to sludge disposal and possible utilization. In addition, the production of sludge from the City's wastewater treatment plants has influenced the disposal alternatives for Pinellas County.   Several regional sludge disposal alternatives have been studied for both entities, as reviewed below.

2.2   PREVIOUS STUDIES

A.   City of St. Petersburg 201 Facilities Plan, 1978 (Reference 10)

The recommended alternative for sludge handling in the 201 Facilities Plan was transportation of dewatered sludge to either the Pinellas County Resource Recovery Facility* or to the County landfill.   This plan was selected based on its flexibility and cost-effectiveness. It was flexible because the alternative would allow the City to negotiate in the future the most cost-effective alternative available for disposal of the sludge cake. Cost-effectiveness of this alternative is derived from its use of the existing thickening and digestion facilities.

The selected sludge disposal alternative required sludge dewatering facilities at each of the City's four wastewater treatment plants. Final land disposal is to be accomplished by one of the following methods:

1.   Transporting dewatered sludge to either the Pinellas County Resource Recovery Facility or the adjacent County-owned and operated landfill, both operational in 1981.   The City can negotiate between these two entities for the least cost alternative.

---

*Ed. Note: The Pinellas County Resource Recovery Facility, still in planning stage, is an incinerator for burning landfill gas and also for grease and sludge, producing steam and electricity.

2.   The pelletizing process at the Largo Wastewater Treatment Plant, and subsequent marketing of the product. The 201 Facilities Plan recommended that the City of St. Petersburg negotiate with the City of Largo as to the cost of processing the dewatered sludge and the quantity of sludge that could be disposed of at the pelletization facility.

B.   Countywide Sludge Disposal Study, 1978
       Chapter 5, Pinellas County 201 Facilities Plan (Reference 11)

This study was written in response to impending Countywide sludge management problems. The disposal methods used by Pinellas County were: 1) liquid land spreading on a rotating field basis, and 2) landfill disposal. Due to stricter state and federal regulations and limited land availability for spreading, these methods are no longer feasible alternatives.

Difficulties in meeting new regulations are partially due to the fact that liquid land spreading practices are difficult to monitor for environmental effects and do not provide 100 percent pathogen kill and pH adjustment to insure the retention of heavy metals.   Landfill disposal of dried and liquid sludge also poses potential problems with respect to groundwater contamination.   While the majority of the sludges originate in the south and central sections of the County, the most suitable land for liquid spreading of sludge is in North Pinellas County or in other counties and is becoming more distant from the sludge producing centers with time.

The recommended plan for disposal of sewage sludge from Pinellas County consisted of four parts as follows:

"First is the accomplishment of sludge concentration and dewatering to a minimum 18 percent dry solids cake at each sludge producing facility in the County. This operation should be the responsibility of the facility producing the sludge.

The second operation is hauling the cake from the individual plants in the County to a central drying facility.   The hauling operation may be conducted by the individual cities involved, by the County, or by a combination of County and City operation so as to assure hauling in the most economical manner.

The third operation should be the heat drying of all sludge produced in the County at a central location near the proposed solid waste incineration facility, from which waste heat from incinerator stack can be utilized to dry sludge.

The fourth and final operation consists of marketing the product produced by the drying operation."

Note that at the present time, each of the City's wastewater treatment plants already has dewatering capability and produces sludge cakes ranging from 15% to 21% dry weight solids.

C.   Summary Report - Landfill Gas Utilization, Pinellas County, Florida, 1978 (Reference 5)

Pinellas County contracted the firm of Henningson, Durham & Richardson, Inc. to study the utilization of landfill gas (LFG) and the disposal of sewage, sludge and grease. In doing so, the County was searching for solutions to 1) odor and gas migration from the Toytown, sod farm, and Bridgeway Acres landfills; 2) the increasing costs and problems associated with the present method of wastewater treatment plant sludge disposal; and 3) grease disposal.

Selection of the recommended plan from the alternative systems considered was based on the following factors:

*   Operating characteristics of
     processes/technologies
*   Materials quantities and characteristics
*   Costs
*   Revenues
*   Interest rates and amortization period
*   Feasibility
*   Implementability
*   Proven/demonstrated technology
*   Funding
*   Environmental Concerns

"All costs are referenced to Engineering News Record (ENR) values for March 31, 1983 (e.g. construction, 4001; skilled labor, 3507; and, materials, 1639. (All costs are presented on a conceptual preliminary design level and therefore have an accuracy of ± 25 percent.

During the system evaluation phase, primary emphasis was placed on economic analysis as the criterion for selection. This was followed closely by proven/demonstrated technology, environmental concerns, and funding. Each system was evaluated at least once. Those appearing to be the most feasible were evaluated at higher levels. The last level of evaluation was economic and was based on a bond analysis where the required bond issue was based on tipping fees calculated representative of "worst" case conditions citing the maximum operating costs with estimated minimum revenues.

The alternative system consisting of sludge and grease incineration; steam boiler, steam (from steam boiler and incinerator waste heat boiler) to Resource Recovery Facility (RRF), with subsequent electricity to Florida Power Corporation (FPC) was found to have the lowest projected tipping cost under worst case conditions; was independent of markets and market values; met the requirements for tax free bond issues (90/10); and had the fewest requirements for backup systems, contractual agreements, and other intangibles."

The recommended alternative includes a high pressure steam boiler, fluidized bed sludge and grease incinerator, grease concentrator, and sludge/grease receiving complex. All project goals were met by the selected alternative: 1) recovered LFG utilization; 2) least cost for sludge and grease disposal, and 3) environmentally acceptable solution.

The economic evaluation of LFG utilization revealed that using steam from a boiler partially fueled by landfill gas could make $600,000 annual revenue. Alternatively, it would cost approximately $101,100 to operate a flare on an annual basis.

D.   Regional Sludge/Grease Disposal Facility - Facility Plan Amendment, Pinellas County, Florida, August, 1984 (Reference 21)

This report was written to partially satisfy the Step 2 facility design analysis for a DER Wastewater Management Grant. It is an amendment to the 1979 201 Facility Plan for Pinellas County.

In the report the following alternative sludge technologies were considered:

* Incineration
* Drying
* Composting

Incineration was selected as the recommended alternative based on the economic analysis. The "Incinerator option" was shown to have the lowest tipping charge to the facility users.

In the report, it was concluded that the recommended Incineration Alternative would be qualified for innovative technology funding. This conclusion was based on the ability of the Pinellas County facility to meet two of the six qualifying criteria for innovative and alternative classification. The two criteria are:

1.   It provides an increased environmental benefit.
2.   It is a net energy producer.

2.3   SUMMARY OF APPLICABLE SLUDGE RULES AND REGULATIONS

A.   Federal Regulations

Recent federal government activities for comprehensive and coordinated control over the disposal and utilization of nonhazardous municipal sludge have been initiated by the passage of the Resource Conservation and Recovery Act of 1976 (RCRA) and the Clean Water Act of 1977 (CWA). As a consequence, the U.S. Environmental Protection Agency (EPA), with contributions from other federal agencies, initiated a program to draft guidelines addressing categories and subcategories of sludge management options. These include:

* Landfilling.

* Surface impoundments (lagoons and stockpiles).

* Landspreading: (a) High rate non-food chain landspreading; (b) non-food chain landspreading; and (c) food chain landspreading.

* Distribution and marketing (the sale or give-away of bagged and bulk sludge-derived products).

* Thermal processing (incineration, pyrolysis, or co-pyrolysis).

* Ocean disposal.

Thus far, the EPA has developed: (1) Interim final rules for land disposal of sludge in landfills and through landspreading; (2) proposed regulations for distribution and marketing (D and M) of sludge; (3) final regulations for control of thermal processing under the Clean Air Act of 1977 (CAA); and (4) the prohibition of ocean disposal under the Marine Protection and Sanctuaries Act of 1972.

Revoking the ocean disposal ban, however, has been considered by the U.S. House of Representatives and Senate committees. These discussions indicate that the EPA may be required to undertake formal procedures to designate ocean sites that are used for dumping. In addition, EPA has acted upon a New York federal district court decision that EPA should not ban ocean dumping without giving equal consideration to the effects of land-based disposal alternatives.

Prior to RCRA and CWA, sludge management was largely the responsibility of the states or individual,private and public entities. Some states regulated sludge management practices indirectly, usually through a variety of broad public health protection powers and other state authorities; other states left sludge management control up to local governments. Federal interest in effective sludge management, however, prompted more states to develop their own programs or to revise existing regulations and guidelines.

Various title 40 code of Federal Regulations (CFR) parts pertain to sludge disposl. Some of these are:

* "PART 240-GUIDELINES FOR THE THERMAL PROCESSING OF SOLID WASTES"

* "PART 241-GUIDELINES FOR THE LAND DISPOSAL OF SOLID WASTES"

* "PART 257-CRITERIA FOR CLASSIFICATION OF SOLID WASTE DISPOSAL AND PRACTICES"

B.   State Regulations

1.   Sludge Classification, Utilization, and Disposal Criteria, FDER 17-7 (Appendix C)

In November, 1983, the Florida Department of Environmental

Regulation (FDER) adopted a rule covering Domestic Sludge Classification, Utilization, and Disposal Criteria. The rule became effective in June 1984 (Part IV, of 17-7 Florida Administrative Code Rule).

The rule applies to domestic wastewater treatment sludges, solid wastes resulting from domestic septage, sewage or food service operations or any other such wastes having similar characteristics. Industrial, air treatment, and water supply treatment sludges are not covered. Criteria and guidance are provided for use of domestic sludges as fertilizer for crops, forage, and as a soil conditioner.

Sludges covered by the rule are classified as Grade I, II and III, with Grade III being of lowest quality. This classification system is based on cadmium, copper, lead, nickel and zinc concentrations. Hazardous waste sludges are regulated under a separate chapter.

The grades of domestic wastewater treatment sludges are shown on Table 2-1.

A sludge is classified as Grade I if concentrations of all of the parameters listed in Table 2-1 are less than the limiting criteria for Grade I sludges, and the sludge is stabilized. In the case of composted domestic sludge, the sludge must also be disinfected.

A sludge is classified as Grade II if concentrations of any of the parameters listed in Table I fall within the criteria for Grade II sludges but no parameters have concentrations above Grade II criteria and the sludge is stabilized and in the case of composted domestic sludge, the sludge is disinfected. A Grade I sludge can be reclassified as a Grade II sludge if the one year moving average of analysis results exceeds the maximum levels for Grade I sludge or the results for any one analysis exceed the maximum Grade I levels by 15 percent.

A sludge is classified as Grade III if concentrations of any of the parameters listed in Table 2-1 are greater than any of the criteria established for Grade III sludges and the sludge is not a hazardous waste as defined by Chapter 17-30, Florida Administrative Code.

Table 2-2 lists the permitting requirements under the new regulations.

2.   Ambient Air Quality (FDER 17-2 Part III)

The drying and combustion sludge treatment processes have gaseous emissions that will require air pollution control devices to meet Ambient Air Quality Standards. The state level requirements for maximum contaminant concentrations for these air pollutants are listed in Table 2-3. Pinellas County has been designated (FDER 17-2.410) as an area not meeting the Ambient Air Quality Standard for ozone (nonattainment area). Therefore, the City of St. Petersburg, located in Pinellas County, is also considered a nonattainment area for ozone. Fortunately, the type

TABLE 2-I

CLASSIFICATION PARAMETERS FOR SLUDGES
STATE OF FLORIDA DEPARTMENT OF ENVIRONMENTAL REGULATION(A)

PARAMETER	I	GRADE II	III
	(Less than or equal to)	Between	(Greater than)
Cadmium	30	30-100	100
Copper	900	900-3000	3000
Lead	1000	1000-1500	1500
Nickel	100	100-500	500
Zinc	1800	1800-10,000	10,000

NOTES:

(a) all values are in mg/kg dry weight.

TABLE 2-2

PERMITTING REQUIREMENTS UNDER
FDER 17-7 PART IV FOR LAND APPLICATION
FAC 17-7.54(4) AND 17-4.64

Conditions to Determine if a Permit is Required

YES - Permit is Required	NO - Permit is Not Required
-Grade II Domestic Sludge not processed	-Grade I Domestic Sludge
-Grade II Composted Domestic Sludge	-Grade I Composted Domestic Sludge
-Any Grade III Sludge	-Stabilized Domestic Septage
	-Stabilized Food Service Sludge
	-Grade I or II "processed" Sludge

TABLE 2-3

STATE OF FLORIDA
AMBIENT AIR QUALITY STANDARDS

Parameter	Time Period	Maximum Allowable Concentration ug/m^3	PPM[a]
Sulfur Dioxide (SO$_2$)	3-hr	1,300	0.5[d]
	24-hr	260	0.1[d]
	Annual mean[b]	60	0.02
Particulate Matter	24-hr	150[d]	
	Annual mean[c]	60	
Carbon Monoxide	1-hr	40,000	35[d]
	8-hr	10,000	9[d]
Ozone (O$_3$)	1-hr	235	0.12[d]

NOTES:

(a)   ug/m^3 = microgram per cubic meter ppm = part per million

(b)   Annual mean - Annual arithmetric mean

(c)   Annual mean - Annual geometric mean

(d)   Not to be exceeded more than once per year

of pollutants found in sludge combustion and drying will not produce ozone, but they will emit gases and odors and require pollution control devices to meet air quality standards.

Besides the standards for Ambient Air Quality, the FDER Chapter 17-2 also lists the maximum allowable increases (prevention of significant deterioration increments) which limits the increases in pollutant concentrations over the baseline concentrations (17-2.310). Also, regulations on gaseous emissions from new sources are addressed under the 7-2.510 Section.

3.   Wastewater facilities (FDER 17-2 Part III)

Although no specific sludge regulations appear in this section, two sections 17-6.090 and 17-6.120 have been reserved for sludge management regulations.   Additionally, there are general regulations that are cited and are applicable to any wastewater treatment facilities under 17-6.040 General Technical Guidance, 17-6.070 Treatment Plants, and Subpart D Compliance.

2.4   EXISTING FACILITIES

A.   General

Sewage collected in the City of St. Petersburg is treated at four (4) City owned and operated plants (WWTP). The wastewater treatment facilities are geographically spaced around the City, one in each corner (see Figure 2-1): Plant #1 (Albert Whitted WWTP), Plant #2 (Northeast WWTP), Plant #3 (Northwest WWTP), and Plant #4 (Southwest WWTP).

The treatment process practiced is completely mixed activated sludge with subsequent anaerobic sludge digestion.   Until June, 1984, dewatered sludge was either hauled to a landfill or used as a fertilizer/soil conditioner at an experimental sod farm. Chlorinated filtered effluent is disposed of through spray irrigational reuse or deep well injection. A summary of the treatment facilities is listed in Table 2-4, and the sludge digestion equipment is listed for each plant in Table 2-5.

B.   Sludge Disposal

Facilities for sludge handling are similar among each of the four plants. The basic system is shown in Figure 2-2. Presently, contract hauling and disposal is the primary method used.   Both the Sod Farm and Toytown Landfill have been closed and are used only in an intermittent manner. After mesophilic (approximately 96°F) anaerobic digestion,the sludge generated at each of the wastewater treatment plants has approximately 2% solids and a liquid consistency. Up to June 1984, this was at the last step in the sludge thickening process train before land application to either the Toytown Landfill, the sod farm, or other landfills in surrounding counties. With the closing of the Toytown Landfill and the sod farm in June of 1984, hauling costs of the liquid

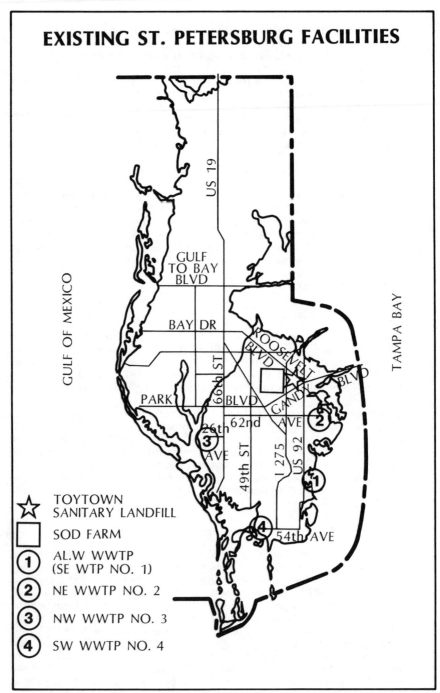

EXISTING ST. PETERSBURG FACILITIES

☆ TOYTOWN
    SANITARY LANDFILL

□ SOD FARM

① AL.W WWTP
    (SE WTP NO. 1)

② NE WWTP NO. 2

③ NW WWTP NO. 3

④ SW WWTP NO. 4

Figure 2-1

TABLE 2-4

SUMMARY OF CITY OF ST. PETERSBURG
TREATMENT FACILITIES

Plant Name	Design Flow (MGD)	Treatment Process[a]
#1 AlbertWhitted	12.4 [b]	CMAS [c]
#2 Northeast	16	CMAS [d]
#3 Northwest	20	CMAS
#4 Southwest	16	CMAS [e]

NOTES:

(a) Treated effluent distributed to users throughout the City for spray irrigation and to injection wells on the wastewater treatment plant site.

(b) 6.6 MGD of wastewater flow will be diverted to the Southwest Plant (#4).

(c) CMAS - Completely Mixed Activated Sludge. Currently being converted from contact stablization.

(d) Modified version of complete-mix activated sludge.

(e) This plant is different from the other three in that it does not employ centrifugation before digestion of the sludge.

TABLE 2-5

EXISTING SLUDGE FACILITIES
AT THE CITY OF ST. PETERSBURG
WASTEWATER TREATMENT PLANTS
#1   ALBERT WHITTED

Process Unit	Description
Anaerobic Digesters	
Number	2
Volume, each, cu. ft.	192,423
Total, cu. ft.	384,845
Belt Filter Presses	
Number	2
Type	Ashbrook-Simon-Hartley Klampresses Size 3, Mark II
Belt Size	2 Meter
Loading (each unit), gpm -Maximum[a] -Percent Solids (range)	160 1 - 4
Discharge [b] -Percent Solids (range)	15 - 21
Design Production (lb. per day dry weight solids)[c]	8,500

NOTES:

(a)   Design capacity is such that if one unit is not functional, the second unit could handle the entire plant sludge feed.

(b)   Minimum cake solids of discharge (% dry basis) depends upon feed solids concentration and is specified by performance guarantee.

(c)   Based on data from Table 3-1 and 17% solids.

TABLE 2-5

EXISTING SLUDGE FACILITIES
AT THE CITY OF ST. PETERSBURG
WASTEWATER TREATMENT PLANTS (Continued)
#2   NORTHEAST

Process Unit	Description
Anaerobic Digesters	
Number	3
Volume	
No. 1, cu. ft.	65,000
No. 2, cu. ft.	65,000
No. 3, cu. ft.	196,350
Total, cu. ft.	326,350
Belt Filter Presses	
Number	2
Type	Ashbrook-Simon-Hartley Klampresses Size 3, Mark II
Belt Size	2 Meter
Loading (each unit), gpm -Maximum[a] -Percent Solids (range)	160 1 - 4
Discharge [b] -Percent Solids (range)	15 - 21
Design Production (lb. per day dry weight solids)[c]	8,600

NOTES:

(a)   Design capacity is such that if one unit is not functional, the second unit could handle the entire plant sludge feed.

(b)   Minimum cake solids of discharge (% dry basis) depends upon feed solids concentration and is specified by performance guarantee.

(c)   Based on data from Table 3-1 and 17% solids.

TABLE 2-5

EXISTING SLUDGE FACILITIES
AT THE CITY OF ST. PETERSBURG
WASTEWATER TREATMENT PLANTS (Continued)
#3   NORTHWEST

Process Unit	Description
<u>Anaerobic Digesters</u>	
Old Plant	
Number	2
Volume, cu. ft. (73,025 each)	146,050
New Plant	
Number	2
Volume, cu. ft. (191,400 each)	382,800
<u>Belt Filter Presses</u>	
Number	2
Type	Ashbrook-Simon-Hartley Klampresses Size 3, Mark II
Belt Size	2 Meter
Loading (each unit), gpm -Maximum[a] -Percent Solids (range)	160 1 - 4
Discharge [b] -Percent Solids (range)	15 - 21
Design Production (lb. per day dry weight solids)[c]	12,900

NOTES:

(a)   Design capacity is such that if one unit is not functional, the second unit could handle the entire plant sludge feed.

(b)   Minimum cake solids of discharge (% dry basis) depends upon feed solids concentration and is specified by performance guarantee.

(c)   Based on data from Table 3-1 and 17% solids.

TABLE 2-5

EXISTING SLUDGE FACILITIES
AT THE CITY OF ST. PETERSBURG
WASTEWATER TREATMENT PLANTS (Continued)
#4   SOUTHWEST

Process Unit	Description
Anaerobic Digesters	
Old Plant	
Number	2
Volume, cu. ft. (200,500 each)	401,000
New Plant	
Number	1
Volume, cu. ft.	174,751
Belt Filter Presses	
Number	2
Type	Ashbrook-Simon-Hartley Klampresses Size 3, Mark II
Belt Size	2 Meter
Loading (each unit), gpm -Maximum[a] -Percent Solids (range)	160 1 - 4
Discharge [b] -Percent Solids (range)	15 - 21
Design Production (lb. per day dry weight solids)[c]	10,000

NOTES:

(a)   Design capacity is such that if one unit is not functional, the second unit could handle the entire plant sludge feed.

(b)   Minimum cake solids of discharge (% dry basis) depends upon feed solids concentration and is specified by performance guarantee.

(c)   Based on data from Table 3-1 and 17% solids.

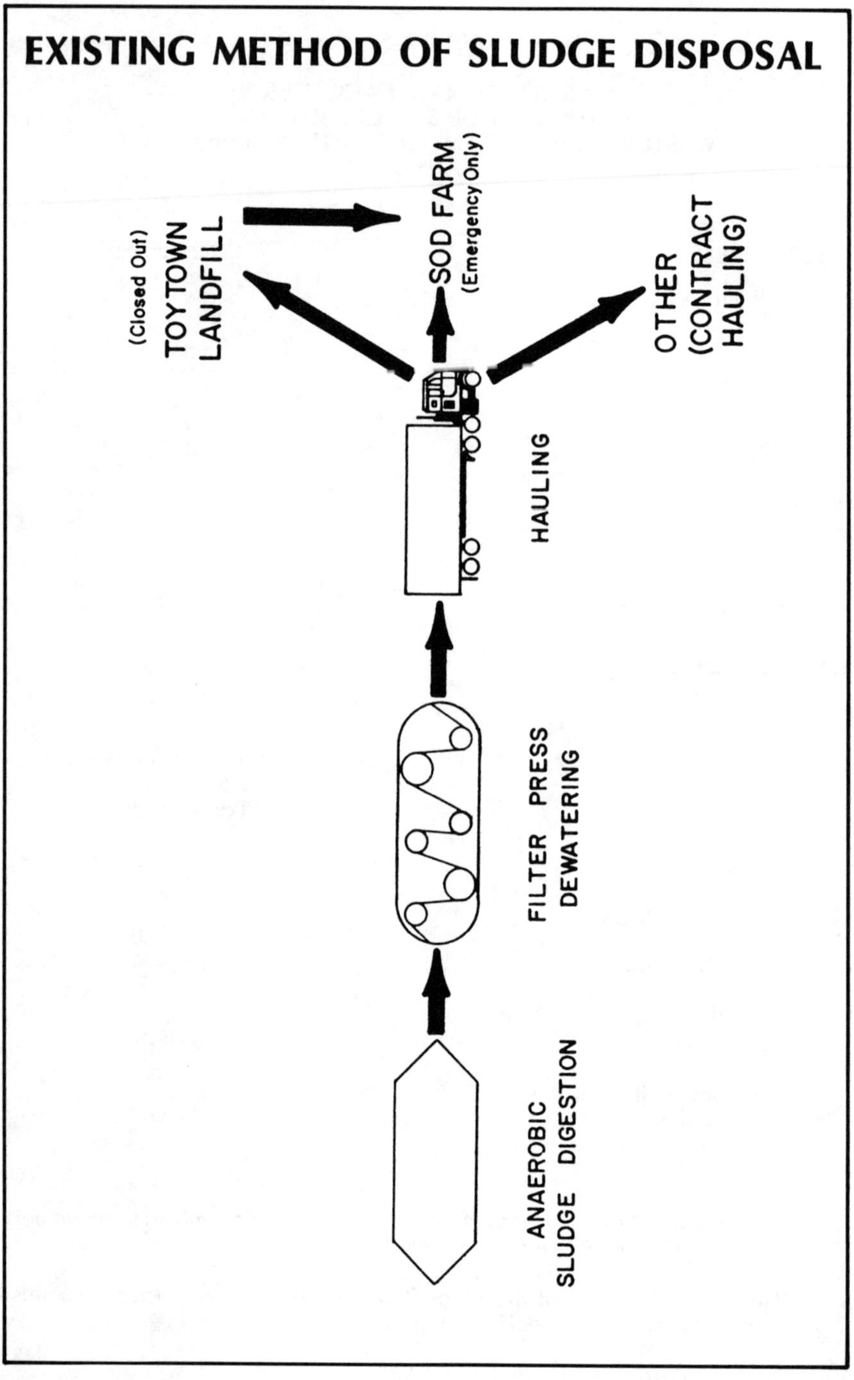
EXISTING METHOD OF SLUDGE DISPOSAL
(Closed Out)
TOYTOWN
LANDFILL
SOD FARM
(Emergency Only)
OTHER
(CONTRACT
HAULING)
HAULING
FILTER PRESS
DEWATERING
ANAEROBIC
SLUDGE DIGESTION

Figure 2-2

sludge would have become excessive. Through a series of studies and evaluations, the City of St. Petersburg decided to purchase belt presses to further dewater the sludge and thereby reduce the hauling costs. In January of 1984 a bid for contract hauling was let out to relieve the City of the task of transporting the sludge to an area landfill. The contract period is until September 30, 1984, with an option for renewal for additional periods of one (1) year each which has been exercised for 1985. The successful bidder provides tractor units and drivers to remove the loaded trailers from the wastewater treatment facility sites and transport them to the approved disposal site(s).

SECTION 3

SLUDGE QUANTITIES AND CHARACTERISTICS

3.1   EXISTING SLUDGE QUANTITIES

A.   Considerations

Numerous factors affect the amount of anaerobically digested sludge produced from each of the four wastewater treatment facilities. Among those factors are the following:

* Maintenance operations on digesters
* Performance of the digesters
* Performance of aeration process
* Performance of centrifuges

In the later part of 1983, operation of the anaerobic digestion systems at all four treatment plants was standardized to include proper heating and primary-secondary digestion. Changes were predicted in supernatant production and digested sludge volume. These changes are significant for the current hauling operation and the impact they will have on running times of the filter presses and on the required capacity of the ultimate disposal alternative.

In the cases of plants #2 and #3, improved operating techniques should give more supernatant and lower wet digested sludge volume. A reduction in wet sludge volume is projected at 10 to 20% in both plants #2 and #3. Plant #1 and #4 have improved operations with the installation of new heat exchange tubes and now produce more digester supernatant than before with less sludge volume.

B.   Dewatered Sludge Production

In Appendix B are tables with raw data for digested sludge produced at each of the four plants. There is an obvious variability in sludge production from month to month due to such factors as listed above in subsection A "Considerations"

The average values for sludge production and their standard deviations are listed in Table 3-1. The amount of digested sludge averaged a total 236,000 gallons per day from all of the four plants.

With the phasing out of the current disposal practice (i.e., hauling of non-dewatered sludge) and the utilization of the belt-filter presses, the amount of sludge to be dealt with will decrease. As shown in Table 3-2, the new volume of sludge should be approximately one-eighth the volume of anaerobically digested liquid sludge. Table 3-2 also shows the volumes of sludge currently produced on a per day basis. If the new belt press sludge dewatering facilities operate at design capacity of 76,800 gallons per day per belt press, then sufficient dewatering equipment has been provided for projected growth. In some instances, such as plant #3, the utilization of the presses will be at nearly 90 percent of capacity. To increase the flexibility of the system the City of St. Petersburg will, if required, provide for hauling the digested sludge to another plant in extenuating circumstances, such as shutting

TABLE 3-1

WASTEWATER TREATMENT PLANT SLUDGE PRODUCTION STATISTICS
(For April 1983 - March 1984)

Plant	Average [a] Flow (MGD)	Digested Sludge Produced [b] (gal/day)		% Solids	
		Mean	Standard Deviation	Mean	Standard Deviation
#1 Southeast	13.8	63,500	20,000	1.6	0.20
#2 Northeast	11.0	45,000	21,500	2.3	0.40
#3 Northwest	14.6	73,500	24,000	2.1	0.25
#4 Southwest	13.5	54,500	22,000	2.2	0.40

81.5

TOTAL SLUDGE PRODUCTION   236,500 gals/day @ 2.0 % Solids

Expected Range 149,000 to 324,000 @ 1.4% to 2.7 % Solids

± 1 std dev

NOTES:

(a)   Plant Influent.

(b)   Based on a 5-day work week.

TABLE 3-2

SLUDGE PRODUCTION
BEFORE AND AFTER BELT FILTER PRESS DEWATERING
(Values based on a 5-day work week)

Plant	Digested [a] Sludge Volume gal/day	% Solids[a]	Daily Pounds-Dry[a] Solids	Daily Dewatered Sludge Production @ 17% Solids		
				Wet Pounds	Gal.[b]	Cu. Yd.[b]
1	63,500	1.6	8,500	50,000	5,700	28
2	45,000	2.3	8,600	50,600	5,800	29
3	73,500	2.1	12,900	75,900	8,600	43
4	54,500	2.2	10,000	58,800	6,700	33
TOTALS	236,500		40,000	235,300	26,800	133

NOTES:

(a)   Based on statistics on sludge production from Table 3-1 and a specific gravity of 1.0. Each belt press has a capacity of 160 gpm or 76,800 gallons per day in an 8-hour day.

(b)   Based on Specific Gravity of 1.05.

down a digester for maintenance. Additionally, a second shift of belt press operation could also be initiated if the volume of sludge exceeds the design capacity.

3.2   EXISTING SLUDGE CHARACTERISTICS

A.   Physical

Anaerobic digestion can result in a 50% reduction of volatile solids which will help to reduce its putresibility and consequently any objectionable odors. With the temperatures required for digestion at about 95°F, some pathogen kill will take place. While the conditions in digesters are unfavorable for the multiplication of most pathogenic organisms, some pathogens will remain viable and the principal bactericidal effect is the natural die-off with long detention times.

Following anaerobic digestion, the next step in all of the City of St. Petersburg wastewater treatment facilities is dewatering. The conditioning of the sludge in anaerobic digestion assists the dewatering process by:

* Improving dewaterability by lowering chemical requirements.
* Reducing the amount of sludge to be dewatered.
* Reducing the odors of the sludge.
* Making the sludge less of a safety hazard with respect to exposure to pathogenic organisms by the operators.

The dewatering treatment itself is useful in further reducing the volume of sludge. A small amount of polymer addition does not significantly change the chemical characteristics of the sludge.

B.   Chemical Characteristics

In November of 1979, a thorough chemical analysis was performed on the various types of samples from the sewage treatment plant, i.e., influent, effuent and sludge. Analyses were conducted for a total of 15 metals and 80 chemicals, resulting in a substantial data base to expand sludge disposal options. These tests demonstrate the absence of any significant contamination. For example, the levels of chlorinated hydrocarbon pesticides such as hexachlorabenzene, tindane and chlordane were below 0.05 mg/kg dry weight in all four wastewater plant sludges. Additionally, to prepare for the new regulations in Part IV of FDER 17-7, the City of St. Petersburg conducted a series of tests (Table 3-3) which could be used to classify their sludge. Sludge at the Southeast Plant #1 was the only Grade II sludge, due to its copper content being greater than 900 mg/kg. Otherwise, the chemical analysis for the St. Petersburg Sludge clearly establishes it as a Grade I sludge from domestic waste.

Typical data on the chemical composition of untreated and digested sludges are reported in Table 3-4. Many of the chemical constituents,

TABLE 3-3[(a)(b)]
SLUDGE TOXICITY ANALYSIS

Constituent	Units[(c)]	Plant #1	Plant #2	Plant #3	Plant #4
pH		7.87	7.60	7.70	7.29
Total Solids	%	2.0	2.1	2.1	1.8
Total Nitrogen	%	8.30	9.26	8.39	7.95
Total Phosphorus	%	2.81	2.89	3.23	3.94
Total Potassium	%	0.56	0.59	0.70	0.50
Cadmium	mg/kg	25	6.1	4.2	6.8
Nickel	mg/kg	140	35	28	41
Lead	mg/kg	600	95	70	170
Zinc	mg/kg	1450	1180	703	350
Copper	mg/kg	2014	640	770	560

NOTES:

(a)   All samples are a composite of four grab samples over a two week period between 2-6-84 and 2-24-84.

(b)   For anaerobically digested sludge before dewatering.

(c)   Milligram per kilogram of dry weight sludge. Percent by dry weight.

TABLE 3-4

TYPICAL CHEMICAL COMPOSITION OF DIGESTED SLUDGE[a]

Item	Digested Sludge	
	Range	Typical
Total dry solids (TS), %	6.0-12.0	10.0
Volatile solids (% of TS)	30-60	40.0
Grease and fats (ether-soluble, % of TS)	5.0-20.0	...
Protein (% of TS)	15-20	18
Nitrogen (N, % of TS)	1.6-6.0	4.0
Phosphorus (P_2O_5, % of TS)	1.5-4.0	2.5
Potash (K_2O, % of TS)	0.0-3.0	1.0
Cellulose (% of TS)	8.0-15.0	10.0
Iron (not as sulfide)	3.0-8.0	4.0
Silica (SiO_2, % of TS)	10.0-20.0	...
pH	6.5-7.5	7.0
Alkalinity (mg/L as $CaCO_3$)	2500-3500	3000
Organic acids (mg/L as HAc)	100-600	200
Thermal content (MJ/kg)[b][c]	6-14	9[d]

NOTES:

(a)   Reference No. 14, from data on both anaerobic and aerobic sludges.

(b)   MJ/kg x 429.92 = Btu/lb

(c)   MJ/KG = millijoules/Kilogram
Btu/lb = Butish thermal units/pound

(d)   Based on 40 percent volatile matter.

including nutrients, are important in considering both the ultimate disposal of the processed sludge and the liquid removed from the sludge during processing. The fertilizer value of sludge, which should be evaluated when the sludge is to be used as a soil conditioner, is based primarily on the content of nitrogen, phosphorus, and potassium.

The thermal content of sludge is important where incineration or some other combustion process is considered, and accurate bomb-calorimeter tests should be conducted so that a heat balance can be made for the combustion system. The thermal content of untreated primary sludge is the highest, especially if it contains appreciable amounts of grease and skimmings. Where kitchen food grinders are used, the volatile and thermal content of the sludge will also be high.

3.3   PROJECTIONS

A.   Projection Considerations

According to the Florida Statistical Abstract 1983, St. Petersburg is the third largest city in Florida with a population of 240,692. During the period of 1980 to 1982 the percent growth in population was only 0.9%. This slow increase is due to the fact that most of the areas of St. Petersburg are already developed. Moreover, there are limited unincorporated areas around St. Petersburg for possible annexation.

The St. Petersburg 201 Facilities Plan (Reference No. 10) presents detailed population and wastewater flow projections to the year 2000. Listed in Table 3-5 are the year 2000 projections from the 201 plan as well as the average flows from 1983. The total flow from the four plants is expected to increase an average of 0.8% over the 17 years from 1983 to 2000 with 53.8 MGD for 1983 to a projected 60.5 MGD for the year 2000. The year 2000 total reflects the inclusion of flows from areas previously served by other wastewater facilities phased out by the 201 plan, such as the Treasure Island WWTP.

B.   Wastewater Treatment Plant Design Capacities

Projected required design capacities for the year 2000 are listed on Table 3-5 for each of the wastewater treatment plants. Changes are being made to the service areas of Plant #1 and Plant #4 which will impact the 201 plan flow estimates for these two wastewater treatment plants. Currently, the influent flow rate at the Plant #1 is around 14.9 MGD, but in the next two years approximately 4.5 MGD will be diverted to Plant #4. The design capacity of Plant #1 will also be revised to approximately 12.4 MGD. This is due to hydraulic modifications that will convert the plant treatment process from contact stabilization to complete mix activated sludge, making its operation similar to the other facilities. Additional changes in the service areas will divert sufficient flow to Plant #4.

C.   Projected Sludge Quantities

TABLE 3-5

COMPARISON OF EXISTING AND
PROJECTED WASTEWATER FLOWS

| Plant | AVERAGE YEARLY FLOW (MGD) | | WWTP (c) Design Capacity F.Y. 2000 |
	1983(a)	2000(b)	
#1 Southeast(d)	13.8	19.2	12.4
#2 Northeast	11.0	14.1	16
#3 Northwest	14.6	13.8	20
#4 Southwest(d)	13.5	13.4	20
TOTAL	52.9	60.5	68.4

Average % Increase per year = 0.80%

NOTES:

(a)   From Table 3-1.

(b)   Projected flows from 201 Facilities Plan (Reference No. 10) (Existing Services Areas).

(c)   WWTP design capacities greater than projected wastewater flow. Flow is to be diverted from Plant #1 to Plant #4 and Plant #1 capacity will be down-rated.

(d)   Flow diversion to Southwest WWTP from Southeast WWTP.

1.   Discussion of Projected Sludge Quantities

In Table 3-6 the present and projected total sludge production from all four of the St. Petersburg wastewater treatment plants are listed. The daily sludge production increases from 20.0 dry tons in the year 1984, to 25.1 dry tons in the year 2007 based upon a five day work week.

The values presented are based on a five-day work week because the belt press dewatering operation is also based on a five-day work week. To convert from five-day to seven-day week, the values listed in Table 3-6 should be multiplied by 5/7.

2.   Assumptions

Projected quantities of dewatered sludge listed in Table 3-6 are based upon several assumptions:

a.   The 1983 average daily sludge production of 20 dry tons (based on a 5 day work week) derived from data over the months of April 1983 to March 1984, represents the present (1984) sludge flow.

b.   Sludge production will increase at an average growth rate of 1% per year, for an overall 17% increase over 17 years.

c.   The dewatered sludge from the filter presses has a solids concentration of 17%.

3.   Discussion

a.   Numerous operational factors will affect the daily, monthly, and yearly averages for sludge production, as discussed in Subsection 3.01A. Therefore, in establishing a base sludge production value, at least 12 months of data should be used. In the case of St. Petersburg, those 12 months should consist of the months after the improvements were made to the anaerobic digestion process in spring of 1983. Consequently the data base for this study is April 1983 to March 1984.

b.   After reviewing the population projections from the Florida Statistical Abstract 1983, and the projected wastewater flows from the 201 Facilities Plan (Reference No. 10), the selected sludge production growth rate of 1% is considered a good estimate. To estimate the growth rate on previous sludge production data may be inappropriate for the following reasons:

*   During the period 1979 to 1982 increased flows and thereby sludge quantities, were due to the incorporation of areas served by wastewater facilities being phased out as part of 201 Facilities Plan.

TABLE 3-6

PROJECTED DAILY SLUDGE PRODUCTION [a]
FOR THE CITY OF ST. PETERSBURG

| Year | Dry Tons/Day | 17% SOLIDS | |
		Cu. Yds/Day[b]	Wet Tons/Day[c]
1984	20.0	133	117.6
1987	20.6	137	121.2
1992	21.6	144	127.0
1997	22.8	152	134.1
2002	23.9	159	140.6
2007	25.1	167	147.6

NOTES:

(a)   Based on a 5 day work week. Values do not include polymer addition.

(b)   Based on volume of sludge = (pounds of dry solids) $\div$ ($C S_S$ x wFs), where $C = 27 \text{ ft}^3/\text{yd}^3$, $S_S = 1.05$, $w = 62.4 \text{ lb/ft}^3$, and $Fs = 0.17$.

(c)   Based on Dry tons $\div$ 0.17 = wet tons.

* A decrease in the quantity of sludge is projected with the improvements to the sludge digestion process making the values for 1983 and beyond slightly lower than what would have been realized otherwise.

c. A 17% solids content of the dewatered sludge is used because:

* This is the percent solids assumed for the incinerator facilities for a possible City/County alternative.

* It is a reasonable mean value for the Ashbrook-Simon-Hartley belt filter presses.

* It is the value used by St. Petersburg in estimating the capacity of their belt presses.

D. Projected Sludge Characteristics

   1. General

   No changes are projected in the physical or chemical characteristics of the St. Petersburg sludge. If industrial waste generators are allowed into the system, they will need to follow the pretreatment standards covered in City Ordinance No. 480-F and No. 703-F (Division 2 and 28 through 51), which are described in detail in Subsection D-2 below. These requirements will control the amount of pollutants allowed in the sewerage system.

   2. Pretreatment Standards

   Ordinance No. 480-F describes requirements for users of the Publicly Owned Treatment Works (POTW's) in the City of St. Petersburg. The ordinance includes the following requirements:

   a. A statement that the quality and quantity of wastewater entering the sewerage system must:

   (1) comply with provisions in the Clean Water Act;
   (2) allow for efficient wastewater treatment; and
   (3) enable the City to meet the effluent limitations on their National Pollutant Discharge Elimination System (NPDES) permit.

   b. A list of prohibited discharges as well as limitations on the chemical and physical characteristics on allowable discharges.

   c. Industrial discharge requirements and criteria.

   d. Criteria for septic tank pumping, hauling and discharging.

   e. Pretreatment requirements.

f.   Requirements for prevention and reduction of accidental discharges.

g.   Documentation such as: a)   plans and specifications, b) hearing procedures and judicial review, and c) penalties for violation.

Ordinance No. 703-F is an amendment to the above Ordinance (No. 480-F) that defines the limitations on effluent concentrations of heavy metals and other by-products from the metal finishing, electroplating, porcelain enameling, electrical and electronic components, and electronic crystal industries. In order to comply with these standards, most industries will be required to pretreat before discharging into the sewerage system.

SECTION 4

AVAILABLE TECHNOLOGY

4.1   INTRODUCTION

A.   General

In response to the growing environmental awareness and dwindling land resources, engineers and scientists are providing new technologies for treatment and disposal of domestic sludge. In this section those technologies are described to develop a rationale for the decisions reached later in the report regarding recommended alternatives. Subsections 4.2, 4.3 and 4.4 discuss post-dewatering sludge treatment processes, while 4.5 covers ultimate disposal methods. Therefore, the processes listed in Subsections 4.2, 4.3 and 4.4 would be used in conjunction with a method of ultimate disposal. Figure 4-1 illustrates the decision-making process in selection of the ultimate sludge disposal method, while the Technology Profiles (Exhibits 4-1 through 4-13) present information on the individual technologies.

B.   Technology Profiles

Technology Profiles are designed to organize a summary of data on each of the available treatment technologies. Some technologies discussed in the following subsections do not have a profile because they are not considered appropriate for the treatment of domestic sludge in St. Petersburg. The technologies not considered for further analyses are:

* Toroidal Drying
* Solvent Extraction Drying
* Spray Drying
* Chemical Fixation
* Co-Composting
* Rotary Kiln
* Wet Air Oxidation
* Pyrolysis/Starved Air Combustion
* Sludge Irrigation
* Sod Farming
* Land Application

The technologies that have been given a profile at the end of this chapter are listed below:

Exhibit 4-1	Flash Drying
Exhibit 4-2	Rotary Drying
Exhibit 4-3	Multiple-effect Drying
Exhibit 4-4	Solar Powered Drying
Exhibit 4-5	Multiple Hearth Incineration
Exhibit 4-6	Fluidized Bed Incineration
Exhibit 4-7	Electric (Infrared) Furnace Incineration
Exhibit 4-8	Co-combustion
Exhibit 4-9	In-vessel Composting
Exhibit 4-10	Windrow Composting

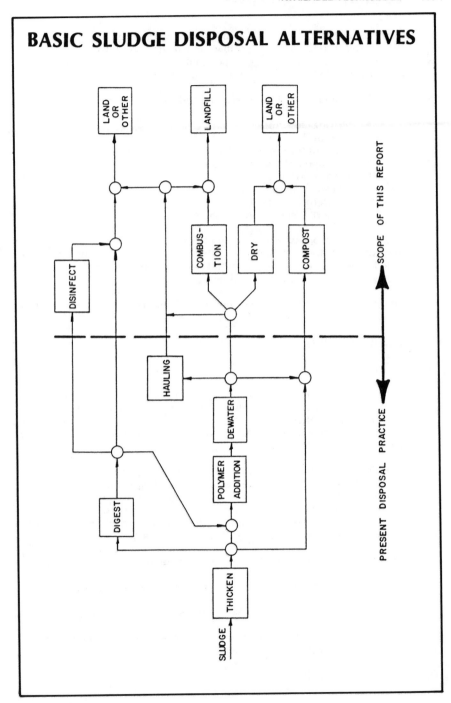

BASIC SLUDGE DISPOSAL ALTERNATIVES

Figure 4-1

Exhibit 4-11   Aerated Pile Composting
Exhibit 4-12   Earthworm Conversion (Vermi Composting)
Exhibit 4-13   Lime Encapsulation

Each Technology Profile presents the following information:

* Brief Description
* Final Product
* Manufacturers/Installations
* Equipment Costs
* Advantages
* Disadvantages
* Potential market
* Operation and Maintenance

A drawing or process schematic is also shown in order to give a general idea of what the technology consists of. The equipment cost range listed indicates only the magnitude of cost for a facility using the technology and is not a cost estimate.

The technologies that are presented in Exhibits 4-1 through 4-13 are reviewed further in Section 5.

4.2   SLUDGE DRYING

A.   General

Even though significant dewatering will be achieved by the belt filter presses, many processes are available that will further dewater sludge and thereby further reduce hauling costs. Generally, sludge drying results in a product high in volatile solids and marketable as a soil conditioner.

The drying technologies listed are heat-drying methods rather than ambient air or composting methods. Composting is covered later under Subsection 4.3. Air drying is not included for the treatment of St. Petersburg sludges due to excessive land requirements, labor intensity, visibility to the general public, and limited additional drying potential for previously dewatered sludge through air drying.

Some environmental effects of drying and combustion processes are:

* Odors and visible emissions require pollution control measures.
* Dust from the drying process is susceptible to fires and explosions.
* Sidestream products are generated and they must be disposed of or treated.

B.   Flash Drying (Exhibit 4-1)

Flash Drying is a process where water in sludge is vaporized very quickly in a stream of hot gas. Basically there are three components to the flash dryer system:

* Sludge Train - wet sludge is mixed in a blending chamber with some previously dried sludge; flash dried by hot gases created by the furnace; and product is separated from gases in a cyclone.
* Furnace - Produces heated air for drying sludge; heat exchanger preheats air with exhaust gases; in some types of equipment partially dried sludge supplements the fuel.
* Induced Draft Facility - consists of air pollution control devices, combustion air heater and the induced draft fan.

The design solids content of a typical dried product is approximately 98 percent. The process requires the addition of ferric chloride ($FeCl_3$) as a conditioner. Some problems encountered are wear on the equipment due to the abrasiveness of the sludge and the accumulation of dust.

C.   Rotary Dryer (Exhibit 4-2)

This type dryer uses a drum-like structure which continuously turns in order to form pelletized sludge. The sloped cylinder moves the material being dried from one end to the other by gravity and forced air draft.

There are two facilities of the direct drying type in Florida: one at Iron Bridge WWTP in Orlando and one at the Largo WWTP. Historical problems associated with these facilities have been:

* Excessive wear on the conveying equipment due to abrasiveness of the sludge.
* Noise pollution.
* Odors.

The end product is a marketable pelletized sludge, that can be sold as soil conditioner or fertilizer.

D.   Toroidal Dryer

A process described in the EPA Process Design Manual (Reference No. 12) is the toroidal dryer. This is a doughnut-shaped dryer which works on a jetmill principle, and, therefore has no moving parts. This process works by transporting the solid material within the drying zone by high-velocity air movement. One of the steps in the process is mechanical dewatering, which insures a 35 to 40 percent solids sludge introduced in the dryer. The process is no longer available.

E.   Solvent Extraction Drying

Solvent extraction, as the name implies, uses a solvent like triethylamine (TEA) to extract water out of a solid. The sludge can be dried at a lower temperature due to the properties of the solvent which reduces energy costs. The solvent is recycled in a completely enclosed system.   The manufacturer states no release of solvent to the environment can occur.   No full-scaled projects have been started utilizing this process.   It is referred to as the B.E.S.T process and marketed by Resource Conservation Company. It is not recommended for further investigation.

F.   Multiple-Effect Drying (Exhibit 4-3)

Multiple-effect drying is an evaporation process which conserves energy by utilizing the vapor generated by the second evaporator (by evaporation of the water from sludge) to dry sludge in the first evaporator. This is referred to as the multiple-effect process and was patented by Carver-Greenfield. Major steps in the process are oil mixing,   multiple-effect   evaporation,   oil-solid   separation,   and condensate-oil separation. The purpose of the oil mixing is to add a fuel oil to the sludge to help prevent scale formation and corrosion of heat exchange elements.

The end products are a useful soil conditioner, sewage oil, and excess steam, all of which may be marketable. The waste products are a condensate high in BOD and gaseous emissions.

The Carver-Greenfield process is widely used in the food and agricultural industries.   Two facilities in Japan treat domestic wastewater sludge and subsequently use the end product for fuel. Some environmental considerations of the process are:

*   Treatment of the vapor condensate to remove ammonia and organics.
*   Treatment of the gaseous emissions by incineration to insure odor destruction.

G.   Solar-Powered Drying (Exhibit 4-4)

Solar-powered drying utilizes solar collection panels to heat air up to 170°F. This heated air is then run through a rotary kiln in a manner similar to rotary drying. The solar powered system operates at a lower temperature than the usual rotary drying temperature of around 1300°F. Therefore, the dried sludge produced will have a higher volatile organic content. Also, the process uses air at a lower velocity and running times are longer, thereby having less problems with dust and emissions.   The only air pollution control device needed is a particulate filtering device.

To date no solar-powered drying facilities exist, although one is currently being built in Sorrento, Florida.   Another solar-powered system that digests or pasteurizes sludge is located in Dade City, Florida. The process for drying is presently being patented, but will be offered to a municipality on a full service operator basis.   The difficulty in applying this technology for a municipality such as the City of St. Petersburg is the land requirement. At least 15 acres of land is required for the equipment for a city the size of St. Petersburg, preferably at the sludge generation sites.

H.   Spray Drying

This is similar to flash drying in that the drying time is very brief. The atomized droplets are usually sprayed downward into a vertical tower through which hot gases pass downward.

There are difficulties with this process. The abrasive nature of the sludge causes the atomizing devices to plug and there are currently no sludge treatment applications available. It is not recommended for further investigation.

4.3   COMBUSTION

A.   General

Combustion processes can also be described as high-temperature oxidation and are useful in situations where "land is scarce, stringent requirements for land disposal exist, destruction of toxic materials is required, or the potential exists for recovery of energy, either with wastewater solids alone or combined with municipal refuse." (Reference No. 12) A summary of combustion processes is found in Table 4-1. A listing of other high temperature processes is found in Table 4-2.

B.   Multiple Hearth Incineration (Exhibit 4-5)

Multiple hearth incineration is described by the EPA as "... durable, relatively simple to operate, and can handle wide fluctuations in feed quality and loading rates..." (Reference No. 12). The multiple hearth and fluid-bed incinerators are the most widely used incinerators in the United States.

The multiple hearth furnace (MHF) is characterized by its series of hearths which represent zones of differing processes. From top to bottom these zones are:

* Drying Zone - evaporation.
* Combustion Zone - elimination of combustible material.
* Fixed Carbon Burning Zone -remaining carbon is oxidized to carbon dioxide.
* Ash Cooling Zone - ash is cooled by the incoming combustion air.

The sequence of the zones for different manufacturers or applications remains the same, but the number of hearths in each zone is dependent upon the quality of the feed, the design of the furnace, and the operational conditions.

TABLE 4-1

COMBUSTION PROCESSES

Mixed municipal refuse technology

Grate-fired (refractory or waterwalled)[a]
    Sludge dried via flue gases
    Sludge dried via steam from furnace
    Sludge added directly to furnace

Vertical packed bed reactors (sludge added to bed)
    Air (Andco-Torrax)
    Oxygen (PUROX^{TM}, a Union Carbide System)

Sludge technology

Multiple-hearth
    Incineration
    Starved-air combustion

Fluid bed

NOTES:

(a)   Same basic process as the Pinellas County Resource Recovery Facility.

TABLE 4-2

OTHER HIGH TEMPERATURE PROCESSES

NAME	TYPE	COMMENT
Wet Air Oxidation	High Pressure/ High Temperature	Primarily used in Industrial Waste Treatment
Reacto-Therm	Starved-Air Combustion	Primarily used in low-volume applications
Modular Starved- Air Incinerators	Two-chambered	No data on sludge incineration alone
Pyro-Sol Process	Pyrolysis	Presently used on refuse waste alone
Bailie Process	Fluid Bed/ Pyrolysis Reactor	Limited full scale data
Wright-Malta Process	Pressurized Rotary Kiln	Limited full scale data
Molten Salt	Pyrolysis	No data on sludge treatment

Problems that have been associated with the MHF system are:

* Failure of the rabble arms and teeth.
* Failure of the hearths.
* Long heating up and cooling down times.
* Odors and unburned hydrocarbons in the exhaust gases requiring afterburners.

Progress is being made to correct these problems, especially with the new metal alloys for the hearth and recycling of exhaust gases for emission control.

C.   Fluidized-bed Incinerator (Exhibit 4-6)

This process consists of a bed of superheated sand into which sludge is introduced. The fluidized-bed furnace (FBF) is relatively simple to operate, has a minimum of mechanical components, and is typically slightly lower in capital cost than the MHF. Strict hydrocarbon emission standards can be met since the exhaust gases are exposed to high temperatures of 1400°F for a few seconds.

There are two basic process configurations for the FBF: 1) In the hot wind box design, the fluidizing air passes through a heat exchanger (recuperator) prior to injection into the combustion chamber. This method increases the thermal efficiency by using the high temperature of the exhaust gases to preheat the incoming air; and 2) In the cold windbox design, the fluidizing air is injected directly into the furnace.

Problems with the FBF have been:

* Feed equipment inadequacies
* Sand scaling on venturi scrubbers
* Erosion from sand bed carry-over

D.   Electric Furnace (Exhibit 4-7)

Electric furnace (EF) is a horizontally oriented rectangular steel shell which imparts electrically derived heat to the sludge moving through it. The EF is suited favorably to a small wastewater treatment operation due to its low capital cost and modular construction. Furthermore, its short start-up and cool-down times make it applicable for intermittent operation.

However, due to the horizontal arrangement, more floor space is required than vertical-type furnaces. Also, replacement of the various components such as the woven-wire belt (3-5 year life) and the infrared heaters (3-year life) are very costly. Finally, the cost of electrical power compared to fossil-fuel power results in greater energy cost.

E.   Single Hearth Cyclonic Furnace

This process sometimes referred to as the single rotary hearth furnace, is a vertical, cylindrical, refractory-lined, steel shelled structure. The cyclonic furnace design differs from the multiple-hearth and fluid bed designs in that it does not allow the combustion air to pass upward through the feed material. The furnace gets the cyclonic description from the tangential injection of the combustion air and supplemental fuel which creates a swirling action.

There are currently 5 facilities in the United States and Canada which either are in operation or in the process of being built. Manufacturers claim that the second generation of cyclonic hearth furnace has eliminated most of the problems associated with past multiple hearth furnaces. One of the major improvements is the ability of the cyclonic furnace to handle autogenous sludges without significant changes in the hearth design and loss of heat value of the sludge. Furthermore, the rotary hearth-type furnace has a relatively low capital cost and is able to meet strict carbonyl or hydrocarbon emission standards. However, this method of incineration is not as accepted in the industry as multiple hearth and fluid bed incineration.

One of the problems associated with the process is a plugging tendency of the fixed plow feeder system.

F.   Pyrolysis/Starved Air Combustion (SAC)

This is the thermal decomposition of sludge in the absence of oxygen or in lower than stoichiometric oxygen atmospheres. The major differences and advantages of pyrolysis over incineration are:

* Production of three potentially useful end products: gas (mostly methane), a liquid, and a char.
* Does not use excess air, hence reducing fan capacities.
* Greater feed capacity (requires less air, so that more sludge can be handled).
* Savings in auxiliary fuels.
* Reduced emissions, since less air is used, less particulates are entrained in the lower air velocities.

Disadvantages are:

* Additional equipment (afterburner and scrubbing).
* Corrosive nature of off-gases.
* Need for additional controls and instrumentation.

This process has proven to be an effective method for burning sludge and meeting strict air quality standards without large amounts of supplemental fuel. However, as with the cyclonic furnace, the process has not been thoroughly tested at full scale. Therefore, it is not as accepted as multiple hearth or fluid bed.

G.   Co-combustion (Exhibit 4-8)

Co-combustion consists of combustion of sludge and solid wastes in the same facility. Currently there are more than twenty sludge and municipal solid waste (MSW) co-combustion systems, including incineration, and pyrolysis/starved-air combustion, that are being operated, tested, or demonstrated in full-scale plants.

There are two basic approaches to co-combustion of sludge with MSW: (a) use of refuse combustion technology by adding dewatered or dried sludge to an MSW combustion unit, and (b) use of sludge combustion technology by adding raw or processed MSW as a supplemental fuel to the sludge furnace.

Co-combustion of sewage sludge with municipal solid waste is a useable approach to solids disposal problems. Not only are both wastes disposed of in an environmentally acceptable manner, but benefits can also be accrued by utilizing the waste heat or combustible exhaust gases. Cost-effectiveness, however, is very site-specific, and in general, co-combustion systems have not been economically feasible without federal and state funding, (Reference No. 12). This is due to the relative costs of disposal and relative quantities of the feed material involved.   For example, solid waste quantities, dry basis, are approximately ten times that of sludge quantities and can be disposed of at one tenth the cost of sludge.

4.4   COMPOSTING AND EARTHWORM CONVERSION

A.   General

Sludge composting is the aerobic decomposition and stabilization of organic constituents to a relatively safe, useful, and aesthetic product by thermophilic organisms. Bacteria, fungi, and actinomycetes are the biological workers that decompose the organic material given the proper volatile organics, moisture, oxygen, carbon/nitrogen ratio, temperature and pH of the substrate.

Compost produced from municipal wastewater sludges can provide a portion of the nutrient requirements for cultivation of plants and helps to condition the soil.   During decomposition of the sludge the temperature can increase to 160°F killing off most of the pathegenic bacteria, eggs, and cysts making it relatively safe for land application.

B.   In-vessel Composting (Exhibit 4-9)

In-vessel composting is also referred to as mechanical composting. Mechanical systems were developed to minimize odors and process time for the composting process by controlling environmental conditions such as air flow, temperature, moisture, oxygen concentration, and volatile solids addition. Carbon content of the substrate is partially controlled by the addition of bulking materials sometimes referred to as carbonaceous material, which can be wood chips, shredded paper, or ground leaves and/or other yard wastes.

Advantages to in-vessel composting are:

* Lower land requirements than unconfined composting.
* Continuous operation.
* Better operation control than unconfined composting.
* Better odor control.

Disadvantages are:

* Higher capital costs.
* More maintenance activities
* Relatively new process for the United States.
* Having to distribute and market end product in order to offset the capital and O & M costs.

Table 4-3 lists some of the available in-vessel composting systems.

C.   Unconfined Composting

In the United States, the most common types of unconfined composting are windrow and aerated static pile.

1.   Windrow (Exhibit 4-10)

The process employs a bulking agent mixed with the dewatered sludge cake to allow natural ventilation with frequent mechanical mixing to maintain aerobic conditions.

In the windrow composting process, the mixture to be composted is stacked in long paralleled rows or windrows. The width of a typical windrow is 15 feet (4.5 m) and the height is 3 to 7 feet (1 to 2m). Based on processing 20 percent solids sludge, land requirements for the windrow process are even greater than for the aerated pile. In an example design problem (Reference No. 12), a 10 MGD facility requires approximately 3 acres, including storage areas and buffer roads.

2.   Aerated Static Pile (Exhibit 4-11)

Aerated static pile was developed to improve upon the windrow system. The forced air method draws in oxygen through the pile. This results in shorter composting times, easier prevention of anaerobic conditions and less risk of odors. Land requirements would be around 2 acres per 10 MGD of wastewater flow.

Advantages to the unconfined composting are:

* Lower capital costs.

* Pathogen kill through natural temperature increase.
* Low additional energy costs and operating expenses.

Disadvantages are:

TABLE 4-3

SOME IN-VESSEL COMPOSTING SYSTEMS
MARKETED IN THE UNITED STATES

NAME	DESCRIPTION
Aerotherm	Circular reactor with staggered augers.
American Bioreactor	Related to BAV process-uses a Tunnel Reactor with no internal moving parts.
American Bio Tech	Uses the Air Lance air distribution system and a vertical reactor.
Fairfield Digester System	Forms a pelletized end-product; limited application in sludge treatment.
Metro-Waste	Marketed by Resource Conversion, Inc.; patented Agriloader System.
Paygro System	Continuous belt in a tunnel chamber.
Purac System	Rectangular reactor is expandible and has an easily accessible parascrew.
Taulman/Weiss	Circular reactor with plug flow through a two-train system.

* Significant land requirement.
* Occasional odor generation.
* Uncertainty in marketing and distribution of end product.

D.   Co-Composting

Municipalities that are having difficulties with disposing of their refuse as well as their sludge, may consider co-composting. In this technology, wastewater treatment sludge is used to add moisture and organic matter to manually and mechanically separated refuse. The composting process takes approximately one week and is in a completely enclosed building. The final product is screened to a 1/4 inch diameter and then used in agriculture, horticulture, landfill top cover, reclamation, and erosion control.

The advantage to co-disposal is usually the reduction of costs for disposal of either refuse or sludge by itself. With this process there may also be an additional revenue from the recyclable materials. However, experience in co-combustion processes has been that the necessary presorting can be costly, time consuming, and inefficient. Another disadvantage to this technology is the lower quality compost produced in comparison to a sludge-only compost. There is lower fertilizer nutrient content in co-disposal compost.

E.   Earthworm Conversion (Exhibit 4-12)

Utilization of earthworms (Vermiculture) for reduction and stabilization of waste products (Vermicomposting) and the resultant production of a viable soil product is not a new concept. It has only been in recent years, however, with the advent of more stringent waste disposal regulations, that vermicomposting has been seriously considered.

The following advantages have been cited by the manufacturers:

* Support facilities are only a series of flat, well drained five foot wide rows, an underdrain collection (if required by regulations), conventional farm type equipment for distribution of feed materials, and a product processing unit (for castings or vermicompost).
* Operating expenses are limited to labor and fuel associated with the distribution of sludge to the rows and the subsequent collection and processing of the castings. No elaborate mechanical type unit processes are required, and there are no delicate or complex instrumentation control loops involved.
* Vermicomposting does not require any complex or dangerous mechanical processes.

A number of potential operating difficulties and their solutions are listed below. The manufacturer claims that these difficulties can be mitigated.

Limitations include:

* The success of a vermicomposting system relies entirely upon the enzymatic efficiency of the worms (Eisenia foetida). Subsequently, as with any biological system, conditions must be kept within the ranges of tolerance with respect to temperature, pH, moisture, oxygen availability, and toxicity.
* To pretreat anaerobically digested sludge so that it is nontoxic to the worms is uneconomic.
* Earthworm conversion decreases the total nitrogen values in the sludge because ammonia nitrogen will be lost to the atmosphere making the product less valuable as a soil conditioner.
* Published information to date is limited on full-scale municipal wastewater treatment plant sludge operations. Consequently, costs are unpredictable.
* Two common ions in municipal wastewater sludge, ammonium and copper, may be toxic to worms. Studies have found that these ions were lethal at additions equivalent to 180 mg NH_4-N and 2,500 mg Cu per kilogram of wet substrate. Safe limits for these elements are not known for earthworms.
* Cadmium accumulates in the worm Eisenia foetida. Zinc apparently does not accumulate in Eisenia foetida but does accumulate in other species. If the worms are to be used as animal feed, the system must be operated such that cadmium and zinc concentrations in the worms do not exceed recommended levels for animal consumption.
* Space requirements may rule out earthworm conversion at some treatment plants.

F.   Lime Encapsulation (Exhibit 4-13)

Encapsulation processes encloses the sludge in a material which is generally impervious. The typical encapsulation processes are expensive and would be employed in disposal of contaminated sludges. These would be (1) the polyethylene process; and (2) the asphalt process, both used in industries with untreatable and dangerous wastes.

Another encapsulation type process, referred to as the lime encapsulation, mixes quicklime (CaO) with dewatered sludge to create a product that is stabilized and disinfected. There are other versions of this process available, but this particular system is reportedly superior for its efficiency. The product from this process is a crumbly and pathogen-free material which can be applied to land as a fertilizer (pH of 12). Design considerations of this process are: (Reference No. 16)

* Minimum amount of additional equipment past dewatering.
* Approximately 1 minute is needed for mixing time.
* 24 hours is needed to reach the temperature and pH for pathogen-kill.
* Lime consumption is 0.27-0.3 kg per kg dry sludge.

* Containers are recommended for storage of product.

This is a relatively new process that would provide an economical method of stabilization and disinfection.

4.5 ULTIMATE SLUDGE DISPOSAL

A. General

With the advent of stricter Federal and State regulations for sludge disposal, the predominant technologies currently used in sludge disposal are different from the ones used 10 years ago. Marine disposal has all but totally been halted in the U.S. and is prohibited under Florida law.

With Part IV of FDER Chapter 17-7, agricultural utilization of unprocessed sludge will be more costly and no longer suitable for all sludges.

An article by J. Pierce and S. Bailey (Reference No. 13) presented the results of a survey conducted in 1982 of sludge disposal practices currently used by Publicly Owned Treatment Works (POTW's) distributed throughout the United States. Below is a summary of the findings of sludge disposal techniques for POTW's with flows greater than or equal to 10 MGD:

* Thermal process was the most common form of sludge disposal (27%)

* Distribution and marketing systems (D&M) for disposal accounted for 22% of the surveyed POTW's

* Other percentages were -

Food-chain landspreading	10%
Non-food-chain landspreading	11%
Landfill	10%
Lagoons or stockpiles	10%
Ocean disposal	4%

* Even though D&M systems are widely used, 80% of the D&M POTW's utilize two or more disposal options

* Half of the D&M POTW's give away their product or sell less than 5% of their sludge

This subsection covers the final disposal methods. Agricultural utilization methods are covered under land applications. Chemical fixation, deep-well injection, and ocean disposal are included in the report for completeness even though they are not considered well suited for the City of St. Petersburg sludge management solutions.

B. Landfill

I.   General

Landfilling may provide an alternative ultimate disposal method for sludge which otherwise would not be suitable for land application due to levels of heavy metals or land availability limitations for additional sludge treatment facilities. For the City of St. Petersburg, the only landfills available are outside the general Tampa Bay area, and contract hauling to the landfill is required. Sludge utilization, either for fertilizer or energy recovery, is not realized through landfilling.

Because the City would not be directly involved in a landfill operation, this subsection deals only with the discussion of what a landfill is and what sludges would be suitable for this method. These factors affect the cost effectiveness of this alternative method.

Landfills are areas set aside for the burying of solid waste, whether it is sludge, refuse, or both. The two principal non-utilization land disposal methods, landfilling and dedicated land disposal, differ in application rates and methods of application. Typical landfill operations involve dewatered sludge subsurface application often at depths of several feet. Dedicated land disposal operations, however, typically involve repetitive liquid sludge applications which may raise the land surface a few inches per year.

2.   Suitability of Sludge for Landfilling

For sludge-only landfills, the solids concentration should be 15 percent or more, so that leaching of nutrients is minimized. In general, only stabilized and dewatered sludges are recommended for landfill disposal. Some of the information required in designing a landfill disposal system would be:

*   Data on average, minimum, and maximum sludge quantities. (Facilities should be sized according to maximum sludge production.)
*   Data on sludge quality which determine what method of landfill disposal can be used and what design levels and types of contaminants should be monitored.

There are no good potential landfill sites in the City of St. Petersburg.

C.   Land Application

I.   General

Land application techniques discussed in this section reflects the utilization of sludge via farming; soil conditioning; and for its use on golf courses, lawns, playgrounds, highway medians, and similar unrestricted public access areas. The EPA has published a process design manual (Reference No. 15) which serves as a guide in site selection, evaluation, and design of municipal sludge land application systems.

All of the City of St. Petersburg's domestic wastewater sludge will continue to be acceptable for land application to some degree under the new FDER Chapter 17-7, Part IV Rules. As with landfill disposal, the City would not be directly involved with the actual operations. Otherwise, a contract-hauler agreement would relieve the City of the responsibility of site selection, permitting, operations, monitoring reports, distribution and marketing.

Land application methods discussed in this subsection will reflect a utilization of the sludge for practical purposes, with the major types of sludge utilization being:

a. Agricultural - sludge is a source of fertilizing nutrients and/or amendments to enhance crop production.
b. Forest - sludge improves forest productivity.
c. Land Reclamation - sludge applied to strip mines, mine tailings, or other disturbed or marginal land, aids the revegetation process and reclamation of these lands.

Agricultural land application is the most likely land application use for the City of St. Petersburg's sludge. There are numerous farms and pasture lands in nearby counties which could benefit from the nutrient content of sludge.

Aiding in the consolidation of phosphate mining gypsum ponds could possibly be another practical use for the sludge. Mixing the sludge with the gypsum in the waste slime ponds may accelerate the evaporation and settling processes, thereby reducing the volume of the pond. The quantity of sludge added would be small compared to the quantity of spent process slurry and would not significantly increase the size of the pond. This is a theoretical disposal method without any installations.

2. Suitability of Sludge for Land Application

Compliance with federal, state and local regulations, is necessary in determining the suitability of the sludge for land application and the rate at which it can be applied. Information about the physical characteristics of the sludge will help in the selection of the transportation and application methods while the chemical and biological characterization will aid in determining:

* the land application option (s)
* appropriate land application rates
* and monitoring parameters

Any future changes in sludge processing and/or characteristics should be considered in designing the selected land application system (s).

A sludge chemical analysis to determine land application suitability should include total N, ammonia N, total P, K, Cu, Zn, Pb, Cd, Ni, Cr, B, As, Al, Co, Mo, Sulfate, PCB's. Also for completeness the levels for other priority pollutants such as halogenated hydrocarbons, aromatic compounds and other organic chemicals should be checked. These tests and more were conducted for the St. Petersburg sludge in 1979, which demonstrated that there were no significant contaminant levels in their sludge. Results from these analyses would generally indicate what contaminants, if any, would possibly pollute groundwater in the area of application or contaminate crops grown on site. Discussed under subsection 2.02B are the new state regulations which have a direct effect on the amount of sludge that can be land applied. For example, under these regulations, sludge from the Northeast WWTP No. I is sometimes classified as Grade II due to its nickel and copper content and therefore is restricted as to where it can be applied and at what rate.

D.   Other

Alternative technologies grouped under this subsection are considered not applicable or practical for a St. Petersburg sludge disposal solution for various reasons. They are included in this study for completeness.

1.   Chemical Fixation

By mixing the sludge with various chemicals such as limestone, flyash, or lime, the chemical fixation process aspires to set the sludge or cement it, into a solid. Cementation materials used in the past are Portland Cements, lime-based mortars, plasters and epoxies. The biggest disadvantage to this process is cost. It is typically used to make somewhat hazardous wastes, such as metal plating, chloralkalies, and flue-gas desulfurization sludge, suitable for landfilling. This would not be appropriate for the relatively clean domestic sludge from St. Petersburg's WWTP's.

2.   Deep Well Injection and Ocean Disposal

a.   Deep well injection of sludge grew from areas where oil drilling rigs were readily available and underground voids left by petroleum extraction needed to be filled. It would neither be cost-effective nor environmentally sound to practice this disposal technique in the City of St. Petersburg.

b.   The two principal methods used in ocean disposal differ by transport mode. The first is by barge which could be old oil tankers or tow barges. The second transport mode is by pipeline which would be long enough to discharge the sludge beyond the continental shelf into the deeper parts of the ocean. In both of these methods, the basic premise is that the sludge will be diluted by the ocean currents and absorbed by the ocean bottom sediments and organisms.

Although this method of disposal is still being practiced in the United States, EPA has begun enacting laws against it. It is not allowed in the State of Florida.

E.   Transportation Considerations

Transport can be a major cost of a land application system. This section is intended to provide a brief summary of the transportation alternatives which may be considered during the preliminary planning phase.

The first consideration is the nature of the sludge itself. As shown in Table 4-4, sewage sludge is classified for handling/transport purposes as either liquid, sludge cake , or dried, depending upon its solids content. Only liquid sludge can be pumped and transported by pipeline. If liquid sludge is transported by truck, rail, or barge, closed vessels must be used. Sludge cake can be transported in watertight trucks, and dry sludge can be transported in open trucks (e.g., dump trucks).

There are four basic modes of sludge transport: truck, pipeline, barge, and railroad. In certain instances, combined transport methods are also used. Some practical considerations of hauling sludge are presented in Table 4-5.

The physical characteristics of interest are solids content, expressed as percent solids. This affects the potential land application system design since:

*   The higher the sludge solids content, the lower the volume of sludge that will have to be transported, stored, etc., because less water must be handled.
*   The type of transport which can be utilized, e.g., truck type, feasibility of pipeline transport, etc.
*   The method of sludge application and sludge application equipment needed, e.g., type of sludge application vehicle, need for incorporating the sludge into the soil, etc.
*   The methods available to transfer and store sludge.

In general, it is less expensive to transport sludge which has a high solids content (dewatered sludge), than sludge with a low solids content (liquid sludge). This cost savings in sludge transport was weighed against the cost of dewatering the sludge by the City of St. Petersburg; the City chose to dewater sludge before transporting it.

Typically, liquid sludge has a solids content of 2 to 10 percent solids, and dewatered sludge has a solids content of 15 to 40 percent solids,

TABLE 4-4

SLUDGE SOLIDS CONTENT AND HANDLING CHARACTERISTICS[a]

Sludge Type	Typical Solids Content (%)	Handling/ Transport Methods
Liquid	1 to 10	Gravity flow, pump, pipeline, tank transport
Sludge cake ("wet" solids)	15 to 30	Conveyor, auger, truck transport (watertight box)
Dried	50 to 95	Conveyor, bucket, truck transport (box)

Notes:

(a)   From Reference No. 15

TABLE 4-5

TRANSPORT MODES FOR SLUDGES

Sludge Type	Transportation Considerations
Liquid Sludge	
Rail Tank Car	100-wet ton (24,000-gal) capacity; suspended solids will settle while in transit.
Barge	Capacity determined by waterway; Chicago has used 1,200-wet ton (290,000-gal) barges. Docking facilities are required.
Pipeline	Need minimum velocity of 1 fps to keep solids in suspension; friction decreases as pipe diameter increases (to the fifth power); buried pipeline suitable for year-round use. High capital costs.
Vehicles Tank Truck	Capacity - up to maximum load allowed on road, usually 6,600 gal. maximum. Can have gravity or pressurized discharge. Field trafficability can be improved by using flotation tires at the cost of rapid tire wear on highways.
Farm Tank Wagon & Tractor	Capacity - 500 to 3,000 gal. Prinicipal use would be field application.
Semisolid or Dried Sludge	
Rail Hopper Car	Need special unloading site and equipment for field application.
Truck	Commercial equipment available to unload and spread on ground; need to level sludge piles if dump truck is used. Spreading can be done by farm manure spreader and tractor.
Farm Manure Spreader	Practical for short haul only.

which includes the chemical additives.   Dried or composted sludge typically has a solids content over 50 percent.

Barging sludge from St. Petersburg to Tampa is no longer a good transportation alternative. The total cost for hauling sludge under this alternative would include both costs for hauling by truck and by barge. In other words it would not be as cost-effective as trucking the sludge by itself.   In a previous report by Initial Engineering (Ref. No. 9), barging was considered a workable disposal method. Several important changes have occurred since then to make this alternative less attractive. These are:

* Liquid sludge is no longer being produced. Therefore, truck hauling costs have been reduced.

* The original plan was to barge the liquid sludge exclusively to a Manatee County Sod Farm near the Tampa Port facility. This sole user is no longer considered available and dewatered sludge will be transported to many areas.

* Liquid sludge was to be pumped from the Albert Whitted (Plant No. 1) directly to the St. Petersburg Port facility. Now the sludge is all dewatered. Dewatered sludge is not as easily pumped.

Exhibit 4-1
FLASH DRYING TECHNOLOGY PROFILE

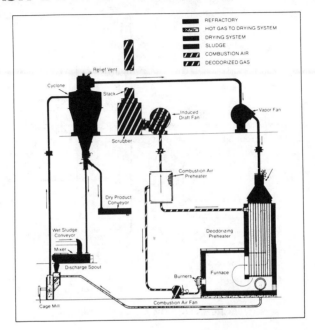

Brief Description:

Drying by furnace and partial pelletization.

Manufacturers:

C-E Raymond/50 installations. Others

Advantages:

o Lower drying temperature than rotary type dryers.
o Nitrogen content is conserved/better marketability.
o Reduction of sludge quantity.
o Typical 98% solids by weight.
o End product is marketable.

Final Product:

Soil conditioner.

Potential Market:

Agriculture or fertilizer industry.

Equipment Costs
(20 Dry tons per day):

$2,000,000 - $2,500,000

Disadvantages:

o Requires fuel to burn off-gases.
o Dust may collect and damage equipment.
o Explosions have occurred with this process.

Operation & Maintenance:

Comparatively complex operation.

Exhibit 4-2
ROTARY DRYING TECHNOLOGY PROFILE

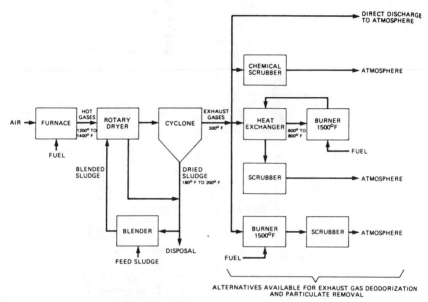

SCHEMATIC FOR A ROTARY DRYER

Brief Description:

Drying in a revolving cyliner.

Manufacturers:

Ecological Services Products/4 installations.
Combustion Engineering/10 installations.
Others

Advantages:

o Two installations in Florida.
o End product is marketable.
o Reduction of sludge quantity 90-95%

Final Product:

Pelletized soil conditioner.

Potential Market:

Agriculture or fertilizer industry.

Capital Costs (20 Dry tons per day):

$4,000,000 - $4,500,000

Disadvantages:

o Excessive wear on equipment due to abrasiveness of the sludge.
o Problems with dust, odors, & noise.
o Energy intensive.
o Sidestream treatment required.
o Product consistency and control.
o Gaseous emissions must be incinerated.

Operation & Maintenance:

Comparatively complex operation.

Exhibit 4-3
MULTIPLE-EFFECT DRYING
TECHNOLOGY PROFILE

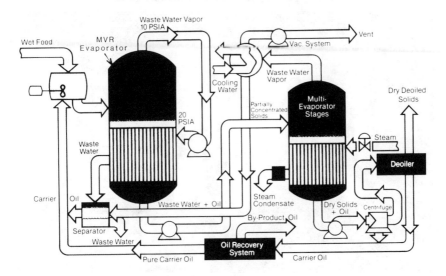

Brief Description:

Drying with reuse of waste steam heat in two stages.

Manufacturers:

Carver-Greenfield/70 installations.

Advantages:

o End product is marketable
o Multiple stage evaporaters use less energy per lb. of sludge than conventional drying.
o Reduction of sludge quantity 95-98% solids.
o Added oil helps to prevent wear on parts.

Final Products:

Soil conditioner and sewage oil.

Potential Markets:

Agriculture or fertilizer industry.

Equipment costs
(20 Dry ton per day):

$3,000,000 - $4,000,000

Disadvantages:

o Complex facility & operations.
o Sidestream contains high concentrations of ammonia and dissolved organics which must be treated.
o Gaseous emissions must be incinerated.
o Patented process-No competition.

Operation & Maintenance:

Comparatively complex operations.

Exhibit 4-4
SOLAR POWERED DRYING TECHNOLOGY PROFILE

JIFFY SOLAR – POWERED SLUDGE TREATMENT FACILITY NEAR DADE CITY

Brief Description:

Sludge is dried by air heated in solar panels.

Manufacturers:

Jiffy Industries. Pilot plant under construction in Sorrento, FL, sludge digester operating in Dade City, FL.

Advantages:

o Power costs are a fraction of the conventional fuel sources.
o Lower operating temperature than rotary dryer.
o End product is marketable.
o Quantity reduction to 92-98% solids.
o Requires minimal air pollution control devices.

Final Product:

Soil Conditioner.

Potential Market:

Agriculture or fertilizer industry.

Equipment Costs (20 Dry tons per day):

$1,300,000 - $1,400,000

Disadvantages:

o Land intensive.
o Relatively new technology.
o Only one manufacturer-single source technology.

Operation & Maintenance:

No long term experience.
No major installations.

Exhibit 4-5
MULTIPLE HEARTH INCINERATION TECHNOLOGY PROFILE

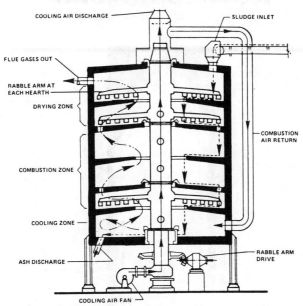

Brief Description:

Combustion of sludge through successive chambers with varying temperatures from 300 to 1,800°F. The cyclonic hearth is a modified multiple hearth.

Manufacturers:

Zimpro
C-E Raymond
Wheelabrator Incineration, Inc.
Nichols
Others

Advantages:

o Minimal land requirement. Most widely used wastewater sludge incinerator in the U.S.(over 350 installations). Waste heat available for power generation.
o Incineration completely deactivates pathogenic organisms.
o Improved operation with mixing jets.
o Maximum reduction of solids.

Final Product:

Ash and potential power generation from waste heat.

Potential Market:

Power Companies.

Equipment Cost (20 Dry tons per day):

$4,000,000 to $6,000,000

Disadvantages:

o Additional fuel needed to burn emissions for odor control.
o Long start-up and cool-down times required.
o High maintenance required for refractory and rabble arms.
o High operation and capital costs.

Operation & Maintenance:

Intensive operational control necessary.

Exhibit 4-6
FLUIDIZED BED INCINERATION TECHNOLOGY PROFILE

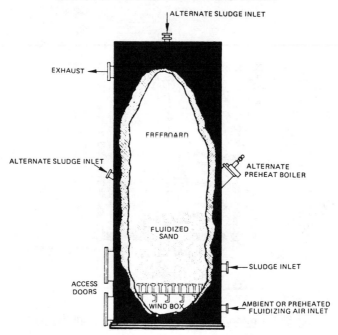

Brief Description:

Vertical cylinder that incinerates sludge in a bed of super heated sand.

Manufacturer:

Dorr-Oliver/200 installations
Drever Company/11 installations
Others

Advantages:

o Minimal land requirement. Rapid start-up (bed retains heat overnight).
o Complete destruction of pathogenic organisms.
o Maximum reduction of solids.
o Slightly lower capital cost than Multiple Hearth.
o Shorter residence times.

Final Product:

Ash and waste heat available for power generation.

Potential Market:

Power Companies.

Equipment Cost (20 Dry tons per day):

$3,000,000 to $4,000,000

Disadvantages:

o Difficult to control temperature in bed and flues.
o Clinker (slag) formation.
o High operation and capital costs.
o Odors may need additional fuel for burning.

Operation & Maintenance:

Comparatively moderate to complex operations.

Exhibit 4-7

ELECTRIC (INFARED) FURNACE INCINERATION TECHNOLOGY PROFILE

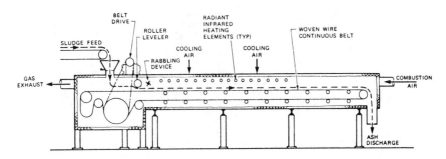

CROSS SECTION OF AN ELECTRIC INFARED FURNACE

Brief Description:

Incineration powered by electricity in a horizontal rectangular steel shell.

Manufacturers/Installations:

Shirco, Inc./13

Advantages:

o Lower capital cost than other incineration processes.
o Completed destruction of pathogens.
o No excess fuel is needed for afterburning.
o Smaller air pollution control devices.

Final Product:

Ash

Potential Market:

None

Equipment Cost (20 Dry tons per day):

$2,000,000 - $3,000,000

Disadvantages:

o Power costs are excessive for the large scale facility needed for St. Petersburg.
o Large capital items need to be replaced after 3 to 5 years.

Operation & Maintenance:

Moderate

Exhibit 4-8

CO-COMBUSTION TECHNOLOGY PROFILE

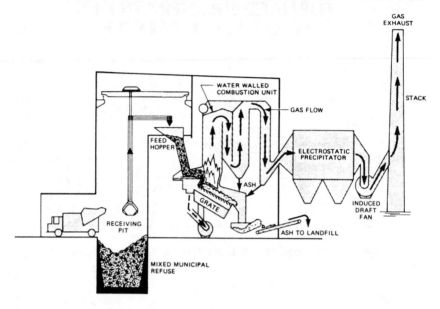

TYPICAL GRATE – FIRED WATERWALLED COMBUSTION UNIT

Brief Description:

Municipal refuse is mixed with sludge before incineration.

Manufacturers:

Industrionics/1 installation in Clayton, GA
C.E-Raymond/installation in Stamford, CT
Others

Advantages:

o Combining of municipal refuse with sludge increases the solids.
o Possibility for resource recovery.
o No market problems.
o Glassy slag helps to physiochemically fix heavy metals.

Final Product:

Ash material and potential power from waste heat.

Potential Market:

Power Companies

Equipment Cost
(20 Dry tons per day):

$5,000,000 to $6,000,000

Disadvantages:

o Additional facilities are required for handling varying feed materials.
o High capital cost.
o Operation is extremely difficult with various feed materials.

Operation & Maintenance:

Relatively complex operation.

Exhibit 4-9

IN-VESSEL COMPOSTING
TECHNOLOGY PROFILE

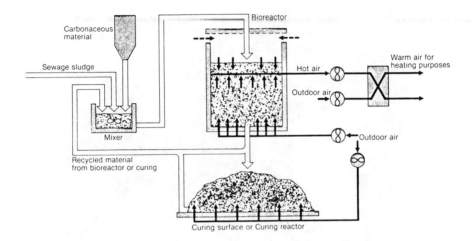

Brief Description and Types:

Sludge treatment in an enclosed building through natural biodegradation.
Types are circular, rectangular, and conveyor systems.

Manufacturers:

Agriloader/10 installations
American Bioreactor/30 installations
American Bio Tech (Airlance)/1 Pilot Study
Fairfield Digestor/8 installations
Purac/800 installations
Metro-Waste/4 installations
Paygro-Compost Systems Co./3 installations
Taulman-Weiss/22 installations
Others

Advantages:

o In-vessel composting eliminates most of the odor and insect problem of unconfined composting.
o Lower land requirement than unconfined composting.

o Continuous operation with better control than unconfined composting.

Final Product:

Soil conditioner or potting soil.

Potential Markets:

Agriculture, fertilizer industry and nurseries.

Equipment Cost
(20 Dry tons per day):

$4,000,000 to $6,000,000

Disadvantages:

o Higher capital cost than unconfined composting.
o Higher energy requirement than unconfined composting.
o Large quantity of end-product.
o No performance record in U.S.

Operation & Maintenance:

Moderate operational requirements.

Exhibit 4-10
WINDROW COMPOSTING
TECHNOLOGY PROFILE

Brief Description:

Composting of sludge on open land with periodic aeration.

Manufacturers*:

Roscoe Brown Corporation
Briscoe-Maphis
Caterpillar
Ag-Chem Equipment Co., inc.
*Equipment Suppliers
Others

Advantages:

o Significant experience.
o Simple treatment process.
o Low capital costs.
o End product is marketable.

Final Product:

Compost used for soil conditioner fertilizer, or potting soil.

Potential Market:

Agriculture, fertilizer industry or nurseries.

Equipment Costs
(20 Dry tons per day):

$150,000 to $200,000

Potential Market:

Agriculture and muck farming.

Equipment Cost
(20 Dry tons per day):

$250,000 to $300,000

Disadvantages:

o Requires considerable labor.
o Requires considerable land.
o Odor and insect problems.
o Large quantity of end product.
o May require disinfection.

Operation & Maintenance:

Comparatively simple operation.

Exhibit 4-11
AERATED PILE COMPOSTING
TECHNOLOGY PROFILE

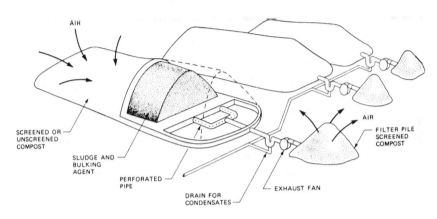

CONFIGURATION OF INDIVIDUAL AERATED PILES

Brief Description:

Stationary piles of compost with
continuous ventilation.

Manufacturers:

5 full-scale testing operations-
o Beltsville, Maryland
o Bangor, Maine
o Durham, New Hampshire
o Detroit, Michigan
o Windsor, Ontario*
* Custom designed from available
  equipment.

Advantages:

o Greater odor control than windrow
  composting.
o Lower land requirement than
  windrow composting.
o Produces sludge product.
o Relatively low capital costs

Final Product:

Compost used for soil conditioner,
fertilizer, and potting soil.

Potential Market:

Agriculture, fertilizer industry and
nurseries.

Equipment Cost
(20 Dry tons per day):

$200,000 to $250,000

Disadvantages:

o Exposed to weather
o Labor, energy intensive.
o Higher operation costs than
  windrow composting.
o Odor and insect problems.
o Large quantity of end product.

Operation & Maintenance:

Comparatively simple.

Exhibit 4-12
EARTHWORM CONVERSION
TECHNOLOGY PROFILE

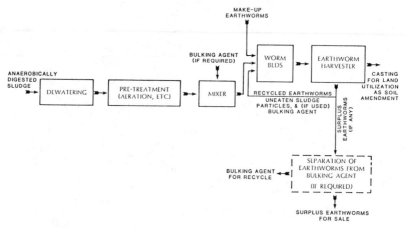

DIAGRAM OF AN EARTHWORM CONVERSION PROCESS

Brief Description:

Uses earthworms to reduce and stabilize sludge.

Manufacturers:

The American Earthworm Company/Pilot Studies in Orlando and Maitland with a recent installation in Kissimmee.

Advantages:

o Relatively low capital cost.
o Relatively low operation and maintenance costs.
o Generally odorless and simple to operate.

Final Product:

Soil conditioner and animal feed.

Potential Market:

Agriculture and animal feed industries.

Equipment (20 Dry tons per day):

$1,800,000 to $2,200,000

Disadvantages:

o Large sidestream productions.
o Decreases total nitrogen content.
o Excessive concentrations of ammonium and copper ions are toxic to worms.
o Requires considerable space.
o May need to pretreat anaerobic sludge.
o Dependent on worm market.

Operation & Maintenance:

Comparatively moderate operations.

Exhibit 4-13
LIME ENCAPSULATION
TECHNOLOGY PROFILE

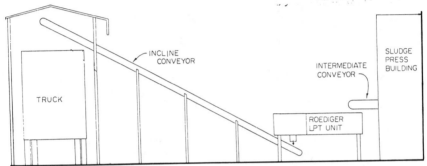

Brief Description:

Quicklime is added to dewatered
sludge to form a crumbly lime-sludge
mixture.

Manufacturers:

Roediger/Pilot studies at Orlando, FL
and Pittsburg, PA and elsewhere in
PA.
Numerous installations in Germany.
Two installations in PA.

Advantages:

o Low capital cost.
o Compatible with belt filter
   presses.
o Makes the sludge more marketable.
o No odors or liquid sidestreams.
o Lime treatment has been
   previously proven effective in
   sludge stabilization.

Final Product:

Lime Encapsulated soil conditioner.

Potential Market:

Agriculture and muck farming.

Equipment Cost
(20 Dry tons per day):

$250,000 to $300,000

Disadvantages:

o 24 hours is needed for complete
   pathogen destruction.
o Substantial chemical costs.
o Single source paddle mixer.

Operation and Maintenance:

Low

SECTION 5

TECHNOLOGY ASSESSMENT

5.1   GENERAL

A.   Considerations

A preliminary screening of the available technologies discussed in Section 4 is performed in this section.   The technologies not excluded by this screening procedure are further analyzed in sections that follow. Using this process, only the practical and reasonably cost-effective solutions are developed in detailed comparative analyses.

In this section the type of technology is reviewed first, i.e., drying or combustion, by a conceptualized screening process using matrix tables. In the summary subsection, technologies selected for further review are then listed with a discussion of the decision-making process.

B.   Screening Criteria

1.   Installations/Proven Technology - is selected as a preliminary screening category due to its importance in establishing the reliability of a sludge treatment or disposal method. Many of the processes listed in Section 4 are relatively new and not yet proven as an effective and economical method. It is not absolutely necessary for one of the newer technologies to meet this criteria for the preliminary screening process. In a St. Petersburg sludge disposal solution, it may be several years before the process would be used allowing time for any current testing or recent installations to prove the merits of the technology.

2.   Compatibility/Other - is a category that screens the technology for compatibility with the existing sludge treatment processes and sludge characteristics.   It also allows for a technology to be excluded on many other bases such as:

        *   Environmental problems
        *   Noise and odors (aesthetics)
        *   Operational problems
        *   Excessive O&M costs
        *   Implementability

3.   Site Availability - is included as a category in the preliminary screening process due to the limited land available for sludge treatment in this densely populated city. Some methods of sludge treatment require large tracts of land within the City. Examples are unconfined composting and solar-powered drying.   These methods would not be feasible with this limitation. However, if these or other disposal methods would utilize land in neighboring counties instead, they would not be excluded as an alternative technology. Because the City does not own a landfill, landfilling is not considered further.

4.   Capital Cost - is a useful screening criteria in assigning an alternative disposal method as either city-owned or privately-owned (full service operator).   Consequently, a technology that

has a relatively high capital cost would not be excluded by this factor alone, but would be reconsidered as a full service operator (FSO) alternative if it qualified as an alternative in the other categories.

5.2   SUMMARY OF SLUDGE DISPOSAL ALTERNATIVES

A.   General

The preliminary screening matrices are presented in Tables 5-1 through 5-5. The column headings listed from left to right are the preliminary screening categories and are discussed above. The technologies are the row headings from top to bottom and include nearly all the processes reviewed in Section 4. Referring to the last column, an "O" represents that the technology is unacceptable or does not meet the criteria, while an "X" indicates that it does fulfill the criteria. In order to be considered as an alternative for sludge management, the technology must be able to receive an "X" in the "acceptability" column.

The preliminary screening process used to choose technologies for sludge disposal alternatives gave the following acceptable alternatives:

* Rotary Drying
* Multiple-effect Drying
* Solar Drying
* In-vessel Composting
* Lime Encapsulation
* Fluidized Bed Incinerator

These six technologies are reviewed in detail in Section 5 as well as Section 8 through transformation curve analysis. Land application is the only ultimate disposal method remaining after the screening process. The only transportation option remaining for further consideration is truck. Each of the above mentioned selected alternatives are considered below.

B.   Rotary Drying

This sludge treatment alternative is considered a candidate for further analysis based upon its relative success in other installations, especially at the City of Largo. In order to be considered further, performance testing for odor and noise controls should be submitted by the manufacturer.

Additionally, if this technology is utilized it must be done by using a full service operation because of the excessive cost to the City. Other things that must be taken into account are potential problems with dust, treatment of sidestreams, and incineration of gaseous emissions. The operation and maintenance cost for this technology is moderately high which will affect any tipping fees charged by a full service operator to the City. The final product, which is a pelletized soil conditioner, could be used agriculturally or as a base for fertilizer.

TABLE 5-1

PRELIMINARY SCREENING MATRIX-

SLUDGE DRYING

Method	Installations/ Proven Technologies	Compat./ Other [a]	Site Avail.	Capital Cost [b]	Acceptability
Flash	X [c]	O	X	O	O
Rotary	X	X	X	O	X [b]
Solvent Extraction	O [d]	X	X	O	
Multiple-Effect	X	X	X	O	X
Solar-Powered	O	X	O	X	X [e]
Spray	O	O	X	O	O

Notes:

(a)   ie - environmental problems, excessive O & M costs, compatibility with existing facilities, and implementability.

(b)   Those technologies that are unacceptable due to excessive capital cost, but otherwise are acceptable will be considered acceptable as a Full Service Operator (FSO) alternative.

(c)   X - Acceptable

(d)   O - Unacceptable

(e)   Acceptability is contingent upon the facility located in a nearby county.

TABLE 5-2

PRELIMINARY SCREENING MATRIX-

COMBUSTION METHODS

Method	Installations/ Proven Technologies	Compat./ Other[a]	Site Avail.	Capital Cost[b]	Accept- ability
Furnaces					
Multiple Hearth	X [c]	O	X	O	O
Fluidized-Bed	X	X	X	O	X [b]
Electric	X	O	X	O	O
Rotary Kiln	O [d]	X	X	O	O
Pyrolysis/Starved Air Combustion	O	O	X	O	O
Co-Incineration	O	O	X	O	O
Wet Air Oxidation	X	O	X	O	O

Notes:

(a)   ie - environmental problems, excessive O & M costs, compatibility with existing facilities, and implementability.

(b)   Those technologies that are unacceptable due to excessive capital cost but otherwise are acceptable, are considered acceptable as a Full Service Operator (FSO) alternative.

(c)   X - Acceptable

(d)   O - Unacceptable

TABLE 5-3

PRELIMINARY SCREENING MATRIX-

COMPOSTING AND OTHERS

Method	Installations/ Proven Technologies	Compat./ Other (a)	Site Avail.	Capital Cost (b)	Accept- ability
In-Vessel	X (c)	X	X	O	X (b)
Unconfined:					
Windrow	X	O	O	X	O
Aerated Static Pile	X	O	O	X	O
Co-Composting	O (d)	O	O	O	O
Earthworm Conversion	O	X	O	X	O
Lime Encapsulation	O	X	X	X	X (e)

Notes:

(a) ie - environmental problems, excessive O & M costs, compatibility with existing facilities, and implementability.

(b) Those technologies that are unacceptable due to excessive capital cost, but otherwise are acceptable, will be considered acceptable as a Full Service Operator (FSO) alternative.

(c) X - Acceptable

(d) O - Unacceptable

(e) Overall acceptability can be proven using pilot scale testing. Alternative warrants further investigation.

TABLE 5-4

PRELIMINARY SCREENING MATRIX-

ULTIMATE SLUDGE DISPOSAL

Method	Installations/ Proven Technologies	Compat./ Other (a)	Site Avail.	Capital Cost (b)	Accept- ability
Landfill	X(c)	O (d)	O	O	O
Land Application	X	X	X	X	X
Deepwell Injection	X	O	X	O	O
Ocean Disposal	X	O (e)	X	O	O
Chemical Fixation	X	O	X	O	O

Notes:

(a) ie - environmental problems, excessive O & M costs, compatibility with existing facilities, and implementability.

(b) Those technologies that are unacceptable due to excessive capital cost but otherwise are acceptable, will be considered acceptable as a Full Service Operator (FSO) alternative.

(c) X - Acceptable

(d) O - Unacceptable

(e) Not allowed by the Florida Department of Environmental Regulation.

TABLE 5-5

PRELIMINARY SCREENING MATRIX-

MODES OF TRANSPORTATION FOR CONTRACT HAULING

Method	Installations/ Proven Technologies	Compat./ Other(a)	Site Avail.	Capital Cost	Accept-ability
Truck	X (b)	X	X	O (c)	X
Barge	X	O	X	O	O
Pipeline	O	O	O	O	O
Railroad	O	O	X	X	O

Notes:

(a)   ie - environmental problems, excessive O & M costs, compatibility with existing facilities, and implementability.

(b)   X - Acceptable

(c)   O - Unacceptable

C.   Multiple-effect Drying (Carver-Greenfield)

Multiple-effect drying is selected as an alternative because of its relative success in the United States in the food and beverage industries. Its lower energy requirement would result in lower tipping fees in a full service operation.

In the preliminary screening, multiple-effect drying had only one criterion that is found to be unacceptable, and this is capital cost. However, this problem can be rectified by going to a full service operation similar to the other technologies which also have high capital costs. Other constraints to be considered are that sidestreams contain very high concentrations of ammonia and dissolved organics which must be treated and gaseous emissions that must be incinerated. The final product could be used as a soil conditioner in the agricultural application or as an auxiliary fuel which could be used in the burners that generate the steam for the process. Operation and maintenance costs are moderate for this technology even with a minimum number of moving parts.

D.   Solar-powered Drying

Solar-powered drying is an available technology which has a low fossil fuel energy usage. It has lower operating temperatures than other processes such as rotary or flash drying. This results in a relative low generation of air pollutants making this technology attractive. In order to use this process for the City of St. Petersburg the facilities must be located where sufficient quantities of land are available to accommodate the space needed for the solar collecters. There is a possibility that sludge can be hauled to another nearby county and treated at a facility where undeveloped land is more readily available. Since power costs are a fraction of the conventional fuel cost, the tipping fee should be relatively low. This is a relatively new technology which in a few years may have proven its reliability and cost-effectiveness.

E.   Fluidized-Bed Incinerator

Of all the composting methods available for sludge incineration, the fluidized-bed incinerator is among the best suited for incineration of domestic wastewater sludge. Additionally, the capital costs on the fluidized bed are generally lower than its nearest competitor, multiple hearth incineration. Thus, the fluidized bed incinerator alternative should have slightly lower tipping fees than the multiple hearth process for the City in the full service operation. Both of the technologies could meet the site availability criteria. The major differences come under the Compatibility/Other criteria, where only fluidized bed and rotary kiln are considered to be either compatible or not having excessive O&M costs or environmental problems. Among the final products, ash and waste heat, waste heat is a recoverable resource available for power generation and subsequent sale to power companies. The heat can also be used in drying sludge in the preliminary treatment for incineration. Before this technology can be recommended, it must be insured by the manufacturer that such problems as temperature control in the bed and flues and clinker formations can be prevented. This technology has, unlike all other combustion methods, a high operation and maintenance requirement.

F.   In-vessel Composting

The composting method selected is the in-vessel type. Because there are numerous in-vessel composting methods, it is not appropriate to exclude any of them at this point in the elimination process. The reasons for selection of in-vessel composting are: reduced land requirement with enclosed composting; and a wide variety of markets for the end-product. The capital cost limitation is avoided by going to the full service operator/tipping fee technique, as is the case with rotary drying and mulitple-effect drying.

Although this technology requires less land than unconfined composting, there is still some on-site storage needed. This may be accomplished through tanks or on concrete slabs under covered areas. The latter would pose potential problems with odors and insects in a humid environment such as St. Petersburg's, and the stored materials may be washed away in heavy rainstorms. Therefore, storage is better accomplished with a separate tank which would result in a slightly higher capital cost for this alternative. Operation and maintenance costs are moderately high. In selection of the type or manufacturer of the In-vessel Composting facilities, the requirement for an easy system to maintain and repair should be made. The final product which can be used as a soil conditioner or as a potting soil has a wide market and a publicly acceptable appearance. Continuous operation with this process allows for better control of composting than with a windrow system and aerated-pile compost system.

G.   Lime Encapsulation

This treatment process is considered feasible because of the manufacturer's claim that it is a low cost method which stabilizes and disinfects the sludge, as well as increases the market appeal of the sludge. It is a relatively new process and will only be used if its cost-effectiveness and reliability can be established.

The disadvantages to this technology are that the on-site storage required in order to allow time for the thermal treatment of sludge for stabilization and disinfection would not be available at most of the treatment plants. Additional capital cost would be incurred in order to build an encapsulated sludge storage facility.   Also, substantial chemical costs are required due to the addition of quicklime for this process. Currently, only one manufacturer can provide the equipment and technical assistance for this technology. However, operation and maintenance costs should be low and the final product will be a marketable commodity, as well as not generating odors or liquid sidestreams that would require further treatment.

This alternative could be funded and operated entirely by the City. A pilot plant study would be recommended in order to determine the reliability and cost effectiveness of this process. With a separate unit required at each treatment plant, the possibility of only constructing units at some of the plants could be investigated where it is necessary to reduce cost of hauling or improve the marketability of the sludge. For example, if sludge at Plant #1 is still a Grade II sludge, this process may improve its classification and, therefore, its marketability.

Of all the technology available for sludge treatment, this has the most potential for being a City-owned and operated facility, mainly due to the low capital cost, relative ease in operation and maintenance, and absence of any odor or liquid sidestream generation.

H.   City and County/Incineration

The technology recommended in this plan and the original report for Pinellas County by Henningson, Durham and Richardson, was the fluidized bed incineration. It was determined in their report that this process could meet the following requirments: utilization of landfill gas (LFG), disposal of sewage sludge, and disposal of grease for the entire county. Although additional capital costs would be incurred in order to accomplish these goals, with grease as a fuel source, recovery of the landfill gas, and federal funding through a state grant recently procured, this alternative becomes cost-effective. Participation by the city would be limited to only utilizing the facility and not contributing capital funds to the construction or design of it.

I.   Land Application/Hauling by Truck

This alternative does not utilize one of the sludge technologies described in Section 4, but utilizes a common means of transportation and the current most widely used means of sludge disposal i.e., land application. Therefore, this alternative seems to be cost effective for the next several years until such time as regulations become more strict, fuel costs become excessive, or land application sites become more scarce. At that time, further sludge treatment may be required and distribution and marketing techniques may be employed.

Land application as a technique for sludge disposal for dewatered sludge would mainly be the utilization of sludge for farming, soil conditioning, and use on restricted public access areas. In order to use the sludge for unrestricted public access areas, it must be treated further for pathogenic organisms.

5.3   STABILIZATION/DISINFECTION CONSIDERATIONS

A.   General

An important area to review in sludge disposal is how well the sludge treatment process will reduce pathogens and improve the aesthetics of the sludge. This subsection reviews what regulations govern these areas of sludge management and how well each technology of the top ten alternatives will meet these regulations.

B.   Regulations

    1.   General

As previously cited in Section 2.3-A. and B., State and Federal regulations have specific guidelines concerning the stabilization/disinfection of sludge. Depending on the final disposal technique used, sludge will require various levels of disinfection. Final disposal methods that would require the least amount of additional stabilization are: ocean dumping, land burial, and subsurface injection. Final disposal methods that would require more treatment for stabilization/disinfection are: distribution and marketing; agricultural land application, and golf course application.

    2.   Minimum pretreatment of sludge before land application should include -Stabilization by means of aerobic or anaerobic treatment to reduce volatile organics by at least forty (40) percent and thereby prevents its odor potential.

    3.   Pathogen Reduction - The stabilized sludge should be subjected to one of the following steps before land application.

        a.   Pasteurization for 30 minutes at 70°C.
        b.   Lime treatment at pH 12.0 for 3 hours.
        c.   Storage for at least 60 days. Storage time in drying beds to achieve less than 60 percent moisture may be considered adequate. The evaporation process causes the reduction of viral infectivity by several orders of magnitude.
        d.   Other methods such as gamma and energized electron radiation, exposure to pathogen reduction. However, very large doses of radiation are required for inactiviation of viruses.

A listing of the various treatment processes used to destroy pathogens and stabilize sewage sludge along with their effectiveness are provided in Table 5-6.

Stabilization and pathogen reduction may not be necessary for sludges subjected to heat drying, composting or chemical treatment.

C.   Review of the top ten alternatives

By applying the information in Table 5-6 (from Ref. No. 20) to the technologies of the top ten alternatives, the performance of each of those alternatives for sludge stabilization/disinfection can be predicted

TABLE 5-6

RELATIVE EFFECTS OF VARIOUS TREATMENT PROCESSES ON DESTRUCTION OF PATHOGENS AND STABILIZATION OF SEWAGE SLUDGES

Processes	Pathogen Reduction	Putrefaction Potential Reduction	Odor Abatement
Anaerobic digestion	Fair	Low	Good
Aerobic digestion	Fair	Low	Good
Chlorination, heavy	Good	Medium	Good
Lime treatment	Good	Medium	Good
Pasteurization (70°C)	Excellent	High	Poor
Ionizing radiation	Excellent	High	Fair
Heat treatment (195°C)	Excellent	High	Poor
Composting (60°C)	Good	Low	Good
Long-term lagooning of digested sludge	Good	---	---
Incineration (500-1,500°C)	Excellent	Excellent	Excellent
High temperature drying	Good	Good	Good

as shown in Table 5-7. Most of the technologies do a good job of reducing pathogens, volatile organics, and odors, except for anaerobic digestion. However, for land application in areas of restricted public access, the present degree of stabilization/disinfection would suffice. Increasing the level of treatment would make the product more marketable.

TABLE 5-7

STABILIZATION/DISINFECTION POTENTIAL
BY THE VARIOUS TECHNOLOGIES

Alternative Technology	Temperature	Process	Overall Potential
Rotary Drying	650°C	High Temp. Drying	Good
Multiple-Effect Drying	250°C	Steam Drying	Good
Solar-Powered Drying	170°C	Pasteurization	Good
In-vessel Composting	60°C	Composting	Good
Lime Encapsulation	70°C	High pH/ Pasteurization	Good
Fluidized-Bed Incineration	500-1500°C	Incineration	Excellent
Land Application of Dewatered Sludge	36°C	Mesophilic Anaerobic Digestion	Fair

SECTION 6

ALTERNATIVE APPROACHES

6.1   INTRODUCTION

In the preceding section, alternative technologies are considered for the ultimate disposal for sanitary sludge from the four (4) regional city facilities. These alternative technologies can be utilized in various approaches to develop and implement a program for the City of St. Petersburg.

In this section, the objectives of the City of St. Petersburg are presented, together with project constraints. Four major non-technical approaches for implementation are presented which provide the structural framework for the subsequent alternative analysis.

    1.   Contracts
    2.   City Owner/Operator
    3.   Regional Incinerator
    4.   Full Service Operator

These alternative approaches should not be confused with the technology specific non-cost factors.

6.2   CITY OF ST. PETERSBURG OBJECTIVES

The City of St. Petersburg objectives can be summarized as follows:

1. **Cost** - The City's objectives regarding cost are to determine a solution which minimizes the initial capital costs along with providing for the lowest possible operation, maintenance, and renewal costs.

2. **Planning Horizons** - The City desires to develop a solution which will serve the City for a minimum of twenty years. If an alternative includes a program of phasing of technologies, then the total program should be evaluated on a twenty year life cycle cost basis.

3. **Reliability** - The recommended program should accept the sludge produced by the City's regional facilities on a continuing basis. Facility redundancy and/or supplemental contracts are to be considered such that the overall program is designed to operate on a continuous basis.

4. **Liability** - The recommended program should minimize the City of St. Petersburg's operational liability for the disposal of sanitary sludges.

5. **Resource Conservation/Reuse** - It is a City objective to reuse the resources available as much as possible. With respect to sanitary sludges, the object is to have a product that is valuable and useful.

6. **Environmental Enhancement** - The environmental quality present in the City of St. Petersburg is to be protected and enhanced, if at all possible.

The above objectives constitute the structural framework and the underlying basis for all subsequent analyses.

6.3   PROJECT CONSTRAINTS

A.   General

Project constraints can be classified in four major areas, including the available methods/technologies, the available financing/capital resources, the existing environmental controls and regulations, and the contract limitations for the investigation. The alternative approaches are designed to overcome the above categories of constraints and provide an implementation mechanism which is useful for the City of St. Petersburg.

B.   Methods and Technologies

With the phasing out of the City's sludge farm operations and the closure of the County's landfill, the existing methods and technologies can not be continued. Sludge dewatering facilities have been provided or are being provided at each plant to minimize the quantity of materials which require transport from each installation. The purchase of additional dewatering facilities allows the consideration of methods or technologies which need dewatering facilities prior to implementation. The selected method must be compatible with the available quality of sludge.

C.   Financing and Capital Resources

Financing and capital constraints are encountered by every entity with respect to ultimate sludge disposal.

Often, solutions have historically included outside private, state, or federal participation due to the usual high initial cost requirements of such solutions. The constraint for this program is to blend the capital requirements and operational requirements into the most acceptable financial package possible for the City of St. Petersburg.

D.   Environmental and Regulatory Constraints

Sanitary sludge is a heterogeneous and time-variable material. Its composition, chemical, and physical characteristics are continuously changing with respect to time, temperature, and various other factors. The recent environmental controls with respect to groundwater pollution, classification of sanitary sludges, and increasingly strict nuisance control measures have resulted in the requirement to process sanitary sludges prior to their ultimate disposal.

Historically, ultimate sludge disposal could be accomplished by the transport and land disposal of the materials in some acceptable location. Current regulations however, especially Chapter 17-7, Part IV of the Florida Administrative Code, require additional facilities, site investigations, and monitoring for land disposal. The recommended alternative must be in compliance with all state and federal regulations.

E.   Contract Limitations

Several assumptions were made prior to the initiation of this investigation. The first assumption was that the City of St. Petersburg would not purchase lands outside the city limits for the use as an ultimate sludge disposal area. The next assumption made was that external sludge processing would be conducted by private entities outside the city limits of the City of St. Petersburg. By blending the capabilities of private enterprise with City facilities, land areas external to the city limits are not required. This contractual constraint has been incorporated into the alternative presented in this report.

6.4   CONTRACT HAULING AND DISPOSAL

A.   General

Typically, hauling and disposal contracts include no permanent capital facilities within the City limits, yet do provide services and management aspects in the ultimate disposal of sludge. The City initially prepares specifications and a typical contract. Then, the City provides a notice to bidders and advertises the procurement. Following the advertisement, the bidders bid the project and the contract is awarded to the lowest responsible bidder. Then the bidder who has become the awardee or contractor prosecutes the assignment during the contracted period. This procedure is straightforward and the City has a significant amount of experience in conducting this type of procurement. Moreover, this is the existing approach to ultimate sludge disposal for the City. In subsequent chapters, the financial ramifications of this approach are discussed.

B.   Existing Contract

The existing sludge hauling and disposal contract and specifications are appended to this report. The contract discusses the following:

    1.   Scope
    2.   Equipment and Personnel
    3.   Contract Time
    4.   Contract Renewal
    5.   Contract Cancellation
    6.   Leases, Licenses, Insurance and Permits

C.   Alternate Forms of Bid and Contract

There are several alternatives available to the City of St. Petersburg with respect to bidding and contracting for services:

Term of Contract - The City can bid alternates varying the term of the contract. Presently the contract provides a one year term renewable for a second year. The City may provide alternate contracting periods. For example, a one year bid term with an annual renewal clause with an escalator for five years; or a two year fixed rate; or a three year term with an escalator; or several types of contracts providing for the best possible terms for a lowest possible bid price. Typically, the longer term contracts with escalator clauses provide an assurance to the City that a hauler/disposer would serve them, with certain recourses available to the City. Moreover, such longer term contracts with escalator clauses provide for a more accurate projection of the annual sludge disposal budget for the Utilities Department. Since the competition in this area, at the present time, has not been intense and the City is located at the same distance from several contracted disposal sites, it may be advisable to procure longer term contracts with price controls in escalator clauses.

Basis of Payment - A second bid alternate can vary the basis of payment. The basis of payment is presently in terms of cubic yards of sludge taken by the contractor. If this basis of payment included a minimum cubic yard production as well as a maximum, then a more competitive bid may be procured. Such minimum and maximum basis of payment with respect to production limits the flexibility of operations for the City, but better defines the contract for the bidder. In our opinion, the City does not pay a premium for the existing basis of payment and therefore the City should not consider a modification of present practice.

Primary and Secondary Contractors - Another alternative is the provision of two awards for two separate contractors. In this manner there is a primary contractor who must perform within a defined limit. If he fails to perform then the City calls a second contractor who has bid on an as-needed basis performance specification and the City can find recourse to the primary contractor for the special performance of the second contractor. The two award process provides some competition between contractors and provides additional assurances that adequate performances will be attained.

Performance Clauses - There are a variety of performance clauses which can be added into contract with haulers and disposers of sanitary sludges. These clauses would provide for such items as punctuality, number of trucks, etc. These performance clauses could be included as necessary in the next hauling contract consideration.

Partial Responsibility Bidding - The City can bid the transport or disposal of the sanitary sludge independently. In this manner, additional contractors may become available in the bidding process. The disadvantage of a haul only and disposal only contract includes additional management and divided responsibilities of the various contractors. The savings in contract price must more than offset the additional difficulties in management of two separate contracts. This opportunity should be compared with the haul and dispose possibilities available to the City. The dispose only contracts have been utilized by various entities in the State. Sod farms around the City of St. Petersburg, as well as the newly formed "approved" ultimate sludge disposal companies are possible dispose only vendors. Due to the restricted availability of appropriate disposal sites in the City of St. Petersburg, the haul only option would not be appropriate, except in conjunction with compatible dispose only contracts.

D.   Advantages

There are numerous advantages to the City contracting approach. Of the many possible advantages, the most noteworthy are as follows:

Flexibility   With the City contracting approach, there are no set technologies utilized. It thereby provides for flexibility in the use of the most competitive technology at that time. Moreover, this approach provides the ability to vary the charge with the volume produced thereby only paying for the amount handled. Other approaches have capacity limitations and there are set costs for not utilizing available capacity. Finally, contracting can be used in conjunction with all other approaches and thereby provides the maximum flexibility possible.

Capital Investment - This approach typically requires little to no capital investment by the City of St. Petersburg. Capital is therefore available for other projects which may be warranted within the City.

Familiarity - The City is very familiar with this approach and there is a historical record within the State of Florida as well as in the City of St. Petersburg with contracting for sanitary sludge disposal.

Responsible Parties - This approach provides for a minimum number of responsible parties in the ultimate disposal of sanitary sludges.

Simplicity - This approach is the simplest implementation system for the City of St. Petersburg. Typically, little effort is required in preparing the documents necessary for contracting and the administration of said contracts.

Competition - There is competition for contracted sludge hauling which provides for the realization of cost effective services. Moreover, presently, the annual costs are as low as any other alternative investigated at this time.

Terms - Contracting provides the ability to modify the terms of the agreement over time to better reflect the needs of the City and to respond to various changes in City policy, environmental regulations, City operations, or other unforeseen events.

E.   Disadvantages

There are several disadvantages in contracting of sanitary sludge management and disposal. Several major disadvantages are as follows:

Reliability - The contracted sludge hauler may not be able to completely serve the City on a continuously reliability basis. Due to this fact, certain entities have been forced to provide for a second contract hauler to serve them in case of non-performance.

Recourse for Non-Performance - There are limited recourses for non-performance in the contracting approach. Typically, replacement of the performance contractor, back billing or withholding payment for services provided are the only recourses available.

Impact of Sludge Rule - It is uncertain what the impact of the new Chapter 17-7, Part IV, Sludge Rule will have upon present practices. Some haulers and sources at DER have indicated that increases in contract prices may result. Others have discussed the limiting of services to a hauling only situation, leaving the responsibility for ultimate disposal to the client or some other entity. The new sludge management rules will most certainly have some impact upon present practices and contracting. The new rules affect the level of confidence which the City may have in contract hauling as a permanent solution.

Price Control - Since this approach has a low capital investment and the present contractors typically have financed facilities on a short term basis, price control opportunities are not provided with this approach. The inflation rate or contracting are expected to equal or approach the inflation rate of commodities in the Tampa Bay area. In contrast, other approaches which include capital investment offer the ability to minimize the impact of inflation and external stimuli, such as new environmental regulations upon the annual price of the services rendered.

Location - The location of the City's sludge generators is not adjacent to the ultimate sludge disposal sites presently being utilized by contractors. The remote source generation provides additional exposure to increases in operating costs.

Dependency on External Entities - In any contracting approach, the City becomes dependent upon the contractor to perform the required services.

Loss of Resource Potential - Once it has been contracted to a private entity for the ultimate disposal of sanitary sludge, the future resource potential is eliminated by contracting away the rights to the materials. This may be overcome by providing appropriate clauses in the contract to share or recover the resource potential of sanitary sludge in the contracting procedure.

Discounted Cost - No state or federal funds are being provided for contract hauling and disposal. Other approaches provide the opportunity for state and federal participation and thereby can be discounted not only on the short term, but also over the life span of the project. The discounting potential of other approaches can reach as much as 75% of the incurred eligible capital costs.

F.   Summary

In summary, the City contracting approach is implementable and presently the only approach utilized. These factors are offset by the uncertainty of the impact of the existing and future sludge rules on present prices, and the reliability of this method of sludge management and disposal without another complimentary sludge management approach. It is noted that this approach is compatible with all other approaches considered and may be an integral part of other approaches utilized. It can also be the primary approach with appropriate contracting clauses and provisions.

6.5   CITY-OWNER/OPERATOR

A.   Description

This approach is the conventional city owned and operated facility. It includes the analysis of the sludge management and disposal problem, the determination of required facilities to correct the problem and provide adequate services, the design of such facilities, the bonding or financing of the facilities, the contracting for construction of the facilities, and finally the start-up and continuous operations of the sludge management and disposal system.

The existing situation provides for contract hauling and disposal of the sludge from the four regional facilities. Since the close-out of the existing landfill and the sale of the sod farm; there are no ongoing operations which may be expanded, modified, or otherwise improved to accommodate the sludge production from the City's facilities. A complete and comprehensive sludge management disposal facilities program would be required.

B.   Advantages

The advantages of the City-Owner/Operator approach include the following:

Operational Control - Of all approaches considered this approach provides the City of St. Petersburg with the greatest operational control of their sludge management and disposal system.

Dependency - This approach minimizes the dependency of the City on others to perform for them the required sludge management and disposal services. Moreover, it simplifies the management of the City's sludge disposal system by providing all services in house.

Reliability - This approach provides for the highest reliability of all the approaches considered. It is assumed that the City of St. Petersburg will continue as a functioning entity, and will be responsible and provide the maintenance, replacement and renewal improvements necessary to attain the highest possible degree of reliability.

City Objectives - With the City-Owner/Operator approach all City objectives can be met with regard to policies, operations, and level of service. There need not be compromise with the desires of other entities with this approach.

Discount Potential - Possible City-Owner/Operator facilities have not been considered this year in the FAC Chapter 17-50 priority ranking State and Federal funding program. However, in the next fiscal year for new projects, it is conceivable that a City-Owner/Operator facility may be granted from 55% to 75% of the eligible project costs, funded through the construction grants program. Our review of the characteristics, rules, regulations, and other matters associated with funding indicate that if the grants program is continued with a sufficient level of funding for next year, such a project is likely to rank in the fundable portion.

Cost Control - This approach provides for the greatest potential of controlling costs and managing the fiscal requirements of system. This is not to say that this approach would result in the lowest total costs, only that the greatest control of costs can be obtained.

City Assets - This approach provides for an increase in the City of St. Petersburg's net assets. This benefit may be quite significant if discounted funds are available, such as State or Federal funding for the project.

Track Record - This approach is the most traditional approach in solving municipal problems and has a long track record of success in the State of Florida.

C.   Disadvantages

The primary disadvantages of the City-Owner/Operator approach are as follows:

Flexibility - This approach has a minimum flexibility when compared to contracting, and/or full service procurement. This approach cannot readily upgrade or significantly modify the selected technology without additional capital requirements.

Initial Capital Requirements - The City Owner/Operator approach provides for the highest initial capital requirements of all the approaches considered. It is necessary to obligate the City's present capital funds for the project. If other more desirable projects for the City are competing for the same capital funds then a selection of the most beneficial projects would have to be made.

Management - This approach requires the greatest increase in staff, liability, degree of management, and other associated activities.

6.6 CITY/COUNTY REGIONAL INCINERATION PROJECT

A.   Description

One of the most popular federally funded approaches in the 1970's included the regionalization of sanitary facilities. The rationale for the regionalization program was that larger facilities could benefit from the economy of scale and that the long term operation and maintenance costs would be reduced in comparison to more numerous smaller facilities. Due to the fact that the initial capital cost for regional systems was seen by the local entities as burdensome, the federal government provided a construction grants program to assist in the implementation of such systems for the public good.

In this investigation, the regionalization approach is limited to the City/County Incineration Alternative. This program is being developed by the County and is presented in detail in Sections 7 and 8.   The regionalization approach is and has been documented as successful when significant State and Federal financial participation is obtained.

B.   Advantages

The major advantages of the City/County Regional Program are as follows:

Defined Program - The City/County Incineration Project has a defined program as a result of the studies conducted for Pinellas County. Moreover, the City of St. Petersburg's 201 Facility Plan stated that a regional sludge disposal alternative would be the best for the area. The new belt filter presses placed in operation at the City's wastewater treatment facilities were provided with the future possibility of transporting the resulting dewatered sludge cake to a regional sludge disposal facility.

Combined Resources - A regional program provides a possibility of combined City and County financial resources as well as the future possibility of incorporating additional entities into the incineration program for the disposal of either sludges or greases.

Economy of Scale - One centralized regional sludge processing facility can attain economy scale in sizing and operations for participating entities.   The incineration technology probably would not be as favorable on an independent entity basis.

Energy Recovery/Landfill Gas Utilization/Waste Disposal - Based upon communications with the City staff, the County has the responsibility of providing for the regional grease disposal. The proposed incineration project is designed to use sludge and grease while utilizing the landfill gas generated from the Toy Town Facilities. The waste heat is used to generate electricity, thus recovering the energy value of the waste products.

Federal Funding - This project has been approved by the Environmental Regulation Commission as eligible for Step 3 construction grants funding and has been ranked as the #5 project in the State. This is the highest priority given to a new project determined to be eligible for funding this fiscal year. The project is eligible for 55% funding of the eligible costs.

Environmental Issues - Several environmental issues have been investigated for this alternative including the possibility of airpollution, water pollution, noise pollution, and solid waste generation. All of the above factors have been determined to be acceptably mitigated or within acceptable regulatory limits.

Solids Disposal - The incineration project generates little solids for the subsequent requirement of marketing, disposal, or additional handling. The ash produced will be disposed of on-site.

Location - The project is located on an industrial area and is surrounded by compatible land uses.

Reliability - The regionalization approach and the technology proposed, both have proven track records.

Capital Costs - In the discussion to date, the City of St. Petersburg does not incur any capital costs associated with this project. The capital costs and local share burden, as well as the ownership of the facilities, are being assumed by Pinellas County.

C.   Disadvantages

The major disadvantages of the City/County Regional Program are as follows:

Assets - The regional approach is intended to be funded by Pinellas County. Therefore no increase in City assets will be attained due to this program. In fact the County assets will derive the benefit of partial Federal funding as well as user contracts supporting the County's debt capacity.

Operations Control - It has been proposed that Pinellas County or its contractee maintain operational control of the facility. The City does not therefore have operational control and the operational flexibility which they may have if the City owned and operated it's own facilities. This disadvantage can be overcome through contractual negotiations to a large extent.

Cost Control - The incineration project is a potentially expensive technology and the operating costs depend significantly upon the cost of electricity which can be produced and the revenue derived therefrom. In addition, it depends on the continuing cheap energy source from the landfill gas generated by the surrounding landfills. If either the revenue stream or gas supply is less than expected, then sufficient increases in cost could result. Since Pinellas County would be the owner of the facility as envisioned, the City is but a large contract user and may not have sufficient control over the future cost of construction and operations. Again this disadvantage can be overcome by appropriate contract negotiations.

Negotiations - This approach requires negotiations between Pinellas County and the City of St. Petersburg to reach an equitable contract protecting both entities' needs with regard to this facility. Moreover, with this alternative other contracts probably will be executed by Pinellas County with other users. The successful contract negotiations and need for continuing contract negotiations are a significant aspect of this alternative.

Technology - The incineration technology was selected for Pinellas County's needs. The City does not require the potentially expensive incineration technology for the combined Grade I sludge. In contrast, considering variable sludge quality from other users as well as the need for the County to dispose of grease, this technology became the most appropriate.

Secondary Benefits - During the contract negotiations, the City must fully investigate the secondary benefits of the ancillary facilities in operational costs at the regional sludge incinerator. Pinellas County will need to provide assurances, costs, and detailed descriptions of the operations such that the City can equitably share in the benefit for the appurtenant systems.

Dependency - This approach requires the City to be dependent upon Pinellas County to perform under the contract terms.

Construction Cost Benefit - Due to the City entering with the County on the regional incineration program, there will be a benefit due to the economy of scale. It has not been decided whether the City will derive the full benefit of economy of scale, or the City and County will share the benefits of this approach.

Federal Funding - The City/County Regional Program approach is dependent upon federal funding. The disadvantages inherent in agency participation also come with the benefits. These disadvantages include the uncertainties regarding amount and timing of grant funds being made available for construction activities as well as the eligibility of the costs incurred for grant funding. Moreover, due to the participation of the USEPA and FDER in the project, construction times and construction costs will probably be greater than those typically encountered in a locally funded project. These increases in costs and delays offset somewhat the advantages of federal funding.

6.7 FULL SERVICE OPERATOR

A. Description

Recently, a new approach to financing sludge improvements has been investigated by entities in the State of Florida. This new approach includes the contracting with a full service operator (FSO) to accomplish the ultimate sludge disposal tasks, and be liable for the applicable rules and regulations associated therewith.

A full service operator may provide complete "turn-key" services including the design, construction, start-up, operation and management of the required facilities or can provide only the contract operations and maintenance of the intended facilities. These two methods are utilized in the private sector approach. Private enterprise participation in the financial aspects of the project are frequently included.

Typically, full service operators have been involved in the "turn-key" provision of the facilities in order to be contracting entities. An example of a full service operator facility would be the Signal-Rescoe facility for resource recovery through the incineration of solid waste for Pinellas County. In this situation the full service operator provides complete management, operations, and facility replacement and renewal as required to maintain contractual obligations with the contracting entity.

B. Procurement Procedures

The full service operator, typically, is procured through a request for proposal and qualifications. Typically, the prospective bidders are asked to submit initially their qualifications and experience and financial capability statement to the contracting entity to be prequalified for the procedure. Prior to prequalification, a packet is provided which includes: an analysis of the problem, a complete definition of the acceptable solution, functional design and description of the desired facilities, legal aspects in contracting, warranties, assurances and idemnification, financial aspects, and engineering and other requirements. The procurement package is distributed to the prequalified bidders. A prebid conference is conducted prior to acceptance of competitive bids. The City evaluates the bids and selects a full service operator to negotiate with, and thereafter implement the project.

C. Advantages

The major advantages of the full service operator approach include the following:

Capital Costs - The City of St. Petersburg can negotiate or determine the extent to which the City desires to participate in the capital costs of the project. This flexibility allows the City to participate to the level they desire from no participation to complete participation.

Privatization - This approach provides for the benefits derived from the rules and regulations concerning privatization, depreciation, bonding advantages and industrial tax incentives. Such tax incentives for the private industry would be incorporated into a lower user fee for the City of St. Petersburg.

Responsibility - The City has the opportunity to structure the amount of responsibility which the City and the full service operator would individually bear, as well as the responsibility which they may decide to jointly share. The level of participation can therefore be optimized from the City's standpoint. Moreover, if the full service operator is obligated to provide the capital as collateral and assurance of performance, then the City may have a very strong recourse for non-performance.

Liability - Through contract, the City may require the full service operator to be liable for obtaining and meeting all local, state, and federal regulations and permit requirements. Moreover, if the full service operator has the capital and operation aspects of the project, then he may be made liable for the appropriate functioning of the equipment, the complete operations on site, and the ultimate disposal of the product or residues derived therefrom.

Operations - Utilizing the full service operator technique, the degree of operational requirements which the City and/or the operator desires to assume, can be optimized with respect to the City's desires. Moreover, the operations can be made flexible within determined bounds to accommodate the operational aspects of the City's sanitary facilities.

Implementation - The full service operator can derive the benefits of a "fast track" implementation through the concurrent scheduling of the preliminary and final design activities as well as the construction activities, start-up activities, and testing of the facilities. Several other entities have benefited from the full service operator technique for the implementation and speed of construction for a project.

D.   Disadvantages

The major disadvantages with this approach include the following:

Track Record - This approach does not have a significant track record for sanitary sludge facilities in the United States.

Assets - Depending upon the participation by the City, the full service operator would have a certain amount of the assets, thereby limiting the asset potential for the City of St. Petersburg. Associated therewith is the potential disadvantage of having the full service operator operating, maintaining and providing replacement and renewal functions for a facility owned, in part, by the City. In contrast, if no City participation in the assets of the facility are provided, the full service operator would be fully responsible for the performance.

Dependency - A full service operator is a private entity and is offered the protection of bankruptcy or other legal means if not performing to the contract requirements. In this aspect the City would have some recourse for the capital assets for the project and may be required to assume the operations of a facility which they previously did not operate, maintain or own.

Contracting - Typically, the full service operator is experienced in contracting these types of facilities. In contrast, the City would have to assure that appropriate safeguards, considerations, and other factors are including in the contract to protect the best interests of the customers of the City.

SECTION 7

DEVELOPMENT OF ALTERNATIVES

7.1   INTRODUCTION

A.   Description of Development

An alternative should include all facets of sludge disposal in order to be an implementable program for sludge management. Figure 7-1 lists all of the available technologies, techniques (approaches), transportation options and ultimate disposal methods addressed in Sections 4 and 6. Each technology that is demonstrated as an acceptable method of further sludge treatment in Section 5 is combined with the best approach from Section 6 and the best method of ultimate disposal to form a workable alternative. After the alternative has been created which satisfies the requirements of a workable program, each of these total sludge management systems is compared in Section 8 in a transformation curve. However, before the transtormation curve analysis can be utilized, costs are calculated for capital, operation, and maintenance. Table 7-1 lists the top ten (10) alternatives with their financial approach, technology and ultimate disposal methodology.

B.   Sludge Management Systems Components

In Subsection 7.2, each alternative is first described by identifying the elements of the alternative. The typical sludge management system flow schematic is as follows:

Belt-pressed   **to** Hauling   **to**   Treatment **to**   Ultimate
Sludge   Facility   Disposal

In some cases the flow schematic will be as follows:

Belt-pressed   **to**   Treatment **to**   Hauling   **to**   Ultimate
Sludge   Facility   Disposal

The components of the system are described as follows:

Belt-pressed Sludge: Each system begins with the sludge produced at the treatment plants throughout the City. The sludge is approximately 17% dry weight solids, and is a product of anaerobic digestion.

Hauling: The sludge is generally transported from each treatment plant to a central sludge treatment facility. This is considered to be internal hauling, ie, between two treatment stages. Contract hauling, as distinguished from internal hauling, is a separate, sludge management system requiring no treatment. The internal hauling for the City is a contractual operation. The City has determined that hauling by City personnel is not cost effective. In the case of an FSO treatment facility, the FSO could also be the hauler.

Treatment Facility: Any of several treatment systems for stabilizing or disposing of the sludge. The ownership of the facility is generally the City or a full service operator (FSO). With FSO ownership, a manufacturer will design, build, operate, finance, and maintain the facility.

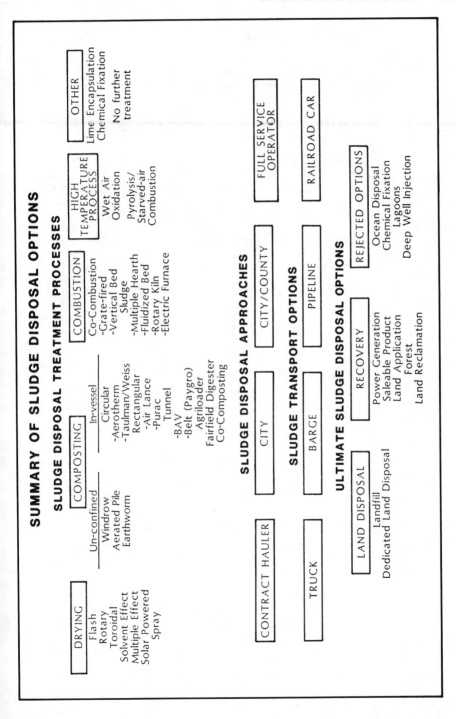

SUMMARY OF SLUDGE DISPOSAL OPTIONS

SLUDGE DISPOSAL TREATMENT PROCESSES

DRYING
Flash
Rotary
Toroidal
Solvent Effect
Multiple Effect
Solar Powered
Spray

COMPOSTING

Un-confined
Windrow
Aerated Pile
Earthworm

In-vessel
Circular
-Aerotherm
-Taulman/Weiss
Rectangular
-Air Lance
-Purac
Tunnel
-BAV
-Belt (Paygro)
Agriloader
Fairfield Digester
Co-Composting

COMBUSTION
Co-Combustion
-Grate-fired
-Vertical Bed
Sludge
-Multiple Hearth
-Fluidized Bed
-Rotary Kiln
-Electric Furnace

HIGH TEMPERATURE PROCESS
Wet Air
Oxidation
Pyrolysis/
Starved-air
Combustion

OTHER
Lime Encapsulation
Chemical Fixation
No further
treatment

SLUDGE DISPOSAL APPROACHES

CONTRACT HAULER

CITY

CITY/COUNTY

FULL SERVICE OPERATOR

SLUDGE TRANSPORT OPTIONS

TRUCK

BARGE

PIPELINE

RAILROAD CAR

ULTIMATE SLUDGE DISPOSAL OPTIONS

LAND DISPOSAL
Landfill
Dedicated Land Disposal

RECOVERY
Power Generation
Saleable Product
Land Application
Forest
Land Reclamation

REJECTED OPTIONS
Ocean Disposal
Chemical Fixation
Lagoons
Deep Well Injection

Figure 7-1

TABLE 7-1

SELECTED ALTERNATIVES FOR
FURTHER COMPARATIVE ANALYSIS

Alt.	Approach	Technology & Disposal Method
A	Full Service Operator	Rotary Drying/D & M [a]
B	Full Service Operator	Multiple-effect Drying/D & M
C	Full Service Operator	Solar Drying/D & M
D	Full Service Operator	Fluidized Bed Incinerator[c]
E	Full Service Operator	In-vessel Composting/D & M
F	Full Service Operator	LimeEncapsulation/Hauling
G	City/County Recovery [c]	Fluidized Bed Incineration Resource
H	City	Lime Encapsulation/Hauling FOB[b]
I	City/Contract	In-Vessel Composting/Hauling FOB
J	Contract Hauling	Truck/Land Application

(a)   D&M - Distribution and Marketing

(b)   Treated sludge made available to haulers free of charge.

(c)   Only ash disposal required.

Ultimate Disposal: Whenever possible, the resultant product is marketable and sold as a soil amendment. In some cases land application of the product is practiced. Distribution and Marketing (D&M) is an important consideration in sludge management when there is a process having such a high capital cost that a substantial revenue is needed to make the alternative cost-effective. This is important in all of the full service operator (FSO) alternatives which base their tipping fees on an expected revenue. The ability to sell the product must be well established before the alternative is recommended.

Backup Management System: It is recommended that with each alternative examined, the City maintain a standby contract for contract hauling. The City is therefore assured of its ability to dispose of sludge through periods of equipment malfunctions, market fluctuations, and during the initial start-up period after construction. Most large municipalities have more than one system of sludge disposal to ensure reliability. Land application for contract hauling includes all methods of sludge utilization such as agricultural, sod farming, land reclamation, and whatever market is currently available to the hauler.

C.   Locational Criteria

The land use plan for the City of St. Petersburg was reviewed for appropriate sites for a centralized sludge treatment facility. There were several major considerations:

-    The facility must not be located in a densely populated area nor should it be surrounded by residential development.

-    The facility should be centrally located, have major traffic arteries, and good connection roads.

-    The facility must not be located in an environmentally sensitive area.

-    Federal regulations require the facility site be above the 100-year flood elevation and further suggest that it be located in an industrially zoned area.

-    To reduce costs to the City, the site should be located on publicly owned property.

The available space at the City-owned wastewater treatment plants was reviewed. The Northeast Plant (No. 2) and the Southwest Plant (No. 4) are the best choice among the 4 wastewater plants for a sludge treatment facility. The main reasons for this selection are: (1) land availability, (2) centralized location, (3) located near main transportation routes, and (4) limited residential development surrounding the plant.

The City owned site south of the sod farm was selected due to the availability of ash disposal facilities and landfill gas for recovery. It is also a site located away from residential areas. The City of St. Petersburg plans to build an industrial park in that area. This would eliminate the use of the site due to the loss in tax revenues for an industrial development. Building an incinerator on this site would also negatively affect the developability of the remaining industrial park site.

Four locations are found to best meet the general criteria:

- Northeast (No. 2) Plant
- Southwest (No. 4) Plant
- Sod Farm
- Toytown Landfill

The final selection depends on the process and final design determinations.

7.2   DESCRIPTION OF ALTERNATIVES

A.   Alternative A - Rotary Drying/Full Service Operator/Distribution and Marketing

The complete sludge management system for this alternative is:

| Belt-pressed Sludge | **to** | Internal Hauling | **to** | Rotary Dryer | **to** | Distribution & Marketing |

The preliminary capital costs for the rotary dryer technology are considered excessive for the City of St. Petersburg. The best way to implement this process would be by a full service operator (FSO). A FSO would charge a tipping fee, set annually by contract. The problem of operating a complex facility would be left up to the FSO. The marketing of the end product would also be the responsibility of the FSO.

The relative success of the rotary dryer at the Largo WWTP and the established market of the end product, makes this alternative an acceptable method for sludge disposal. The Tampa Bay area has several fertilizer companies that would possibly buy the dried sludge. In fact, the City of Houston, Texas is selling all of its flash dried sludge to a fertilizer company in Tampa.

Additionally, this alternative has had nuisance difficulties at both the Largo and Iron Bridge (Orlando) facilities. At the Largo plant, the complaints are about odor. At the Iron Bridge plant, the complaints are about odor and noise. In both cases, it has caused the operators to temporarily shut-down the process.

B.   Alternative B - Multiple-effect Drying (Carver-Greenfield)/Full Service Operator/Distribution and Marketing

The complete sludge management system for this alternative is:

| Belt-pressed | **to** | Internal | **to** | Drying | **to** | Distribution |
| Sludge | | Hauling | | Facility | | & Marketing |

The sludge treatment process refered to as multiple-effect drying, was selected for additional comparative analysis mainly due to its energy efficiency and proven reliability. As with any high-tech process, the capital cost is relatively high and the process requires close surveillance. Therefore, the preferable management for this technology is by full service operator (FSO). A FSO operation will relieve the City of the tasks to design and build the facility, operate and maintain the process, and dispose of the end product.

There are additional considerations in this analysis which will increase the desirability of full service operation. Other end products, such as sewage oil and excess steam may be marketed by the FSO company. The FSO could accept scum and grease from the St. Petersburg plant for the production of sewage oil. This process is also amendable to septage drying, but the treatment of the resulting sidestream biological oxygen demand (BOD), must be considered. This sidestream would be recirculated through the wastewater treatment plant and the cost of treatment would be incurred by the.Full Service Operator.

C.   Alternative C - Solar Drying/Full Service Operator/Distribution and Marketing

The complete sludge handling system in this alternative is:

| Belt-pressed | **to** | FSO | **to** | Solar Drying | **to** | Distribution |
| Sludge | | Hauling | | Facility | | & Marketing |

Of all the technologies listed in Table 7-1, solar drying is the most recent technology in the sludge treatment industry. Solar drying has been previously used in the form of sludge drying beds. However, this technology now provides a method of collecting solar heat and using it more effectively in drying sludge. Not only is the final product more marketable and easier to handle, but the facilities are not as affected by rainstorms and flooding, as would the drying beds. Odor and insect problems are also minimized.

Because the technology has only one installation and cannot be considered a proven method, the best plan for utilization would be the Full Service Operator. The manufacturer, reportedly was agreeable to this arrangement and has a plan of accepting St. Petersburg sludge in limited quantities until the cost-effectiveness and reliability of the technology can be demonstrated to the City. In fact, at the time of this writing, the patent for the process has not been obtained, and the manufacturer is not willing to sell the equipment to the City until the patent has been obtained.

For the alternative analysis, the capital and O&M costs were estimated based on the proposal of a Hillsborough County site. The City would pay the labor and fuel costs for a hauler to deliver sludge to this facility. Then the full service operator would treat, distribute, and market the sludge and charge the City a tipping fee accordingly.

D.   Alternative D - Fluidized Bed Incinerator/Full Service Operator
      The complete sludge management system for this alternative is:

Belt-pressed	to	Internal	to	Fluidized Bed	to	Ash
Sludge		Hauling		Incinerator		Disposal

The technology, fluidized bed incineration, used in this alternative is the same as the City/County Alternative G. However, Alternative D serves only the City by a full service operator, while Alternative G serves the entire County and is owned by the County.

The disadvantages to having the City-only facilities would be:

-    no economy of scale with smaller facility

-    a much lesser grease input at the City Facility, therefore, more auxiliary fuel required

The advantages to a City-only incinerator alternative would be:

-    slightly reduced hauling costs over other City-only alternatives due to no internal hauling between wastewater treatment plants

E.   Alternative E - In-Vessel Composting/Full Service Operator/ Distribution and Marketing

The complete sludge handling system in this alternative is:

Belt-pressed	to	Internal	to	In-vessel	to	Distribution
Sludge		Hauling		Composting		& Marketing

Of all the composting methods available, (windrow, static aerated pile, and in-vessel composting) the in-vessel composting technique was determined to be best suited for the City of St. Petersburg. However, due to the high capital cost, it is best implemented as a full service operation. As with the other alternatives, a contract hauling agreement should be maintained to dispose of sludge (liquid, dewatered, or composted) when one of the treatment processes does not do its job.

The end product of this process produces a marketable compost as demonstrated in a report titled "A Market Utilization Study for Municipal Sludge - Compost South Florida" by the firm, The Matrix Company.

The main advantages of a composted sludge in distribution and marketing are:

-    material adds humus and the composted sludge has the

consistency of peat.   Peat has a well established market.

- with the new Florida DER 17-7 Part IV regulation, a sludge that is properly composted can have a non-restricted use.

Some disadvantages to composting would be:

- addition of bulking material to sludge will increase the operation and hauling costs due to the increase material generated.

Finally, in-vessel composting is considered an acceptable treatment technology due to its success in Germany, its limited application in the United States, ability to meet DER disinfection requirements, and marketability of the end product.

F.   Alternative F - Lime Encapsulation/Full Service Operator/Hauling

The complete sludge management system for this alternative is:

| Belt-pressed Sludge | **to** | Lime Encapsulation | **to** | Contract Hauling | **to** | Land Application |

Lime encapsulation is a technology that has been demonstrated in Europe but, like in-vessel composting and solar drying, has yet to be established as a proven technology in Florida. Because of this status, the full service operator arrangement may prove cost-effective and allow protection for the City in the case where the system would not be successful.   However, capital costs and operation and maintenance costs appear to be very low, making a City-owned facility feasible. A pilot study or a single unit installation would test the process and provide a basis for determination.   As with the other processes, Contract Hauling will be maintained, and in the worst case would be the method of disposing of the lime encapsulated sludge.   The best case situation is to sell or give the product away by making the product free to haulers Freight on Board (FOB).   The bagged product could be made available to the public, if the demand exists for it.

G.   Alternative G - Fluidized Bed Incineration/Resource Recovery/City and County

The complete sludge handling system for this alternative is:

| Belt-pressed Sludge | **to** | Internal Hauling | **to** | Pinellas County Sludge Incinerator |

Pinellas County is currently pursuing a plan to design, build, and operate a fluidized bed incinerator to serve the County and the City. The City's involvement is limited to use of the facilty and not funding or ownership.   There was a good possibility that the facility will be partially funded with a state grant.

For the County, the incinerator was found to be the best available technology for sludge and grease disposal by the County consultants. The project will accomplish most of the desired goals, including:

- resource recovery through utilization of landfill gas as a fuel source
- effective disposal of grease
- resource recovery of waste heat through generation of electricity

The City hauls the dewatered sludge to the facility. Therefore, this cost is added to total cost of this alternative in Subsection 7.4. And as before, Contract Hauling will be maintained in case of operational problems at the incinerator.

H.   Alternative H - Lime Encapsulation/City-Owned

The complete sludge management system for this alternative is:

Belt-pressed **to** Contract   **to** Land
Sludge   Hauling   Application

This alternative is similar to Alternative F - Lime Encapsulation/Full Service Operator. In this arrangement, the City as owner will benefit from possible grant funding and the tax advantages of a capital asset, but will have the responsibility of disposing the product.

In the worst case, the disposal of lime encapsulated sludge would cost similar to the current hauling. In the best case, the product would be sold for a modest price. The middle alternative is to make the treated sludge to haulers free of charge.

The capital cost of this alternative would include:

- Mixer equipment
- Lime silo
- Moveable Storage bins
- Covered storage areas

Operation costs are minimal due to low energy usage and the utilization of the belt-press operator for running the simple system.

Advantages of the system are:

- increased salability of sludge due to increased pH, disinfection, and stability
- simple operation and minimal maintenance
- low capital cost
- significant stabilization
- no sidestream production

Disadvantages of the system are:

- requirement of lime feed and therefore increased

chemical costs
- potential odors from chemical reactions
- limited percent solids without further drying

I.   Alternative I - In-Vessel Composting/City Owned/Hauling

The complete sludge handling system for this alternative is:

Belt-pressed   **to**   Internal   **to** In-vessel   **to** Distribution
Sludge   Hauling   Composting   & Marketing

It is possibile that the City can obtain grant funding for composting, therefore a City-owned in-vessel composting facility has been selected as an alternative for further analysis. In-vessel composting is more complex than windrow, but still can be easily managed by City personnel. The difficulty will be in marketing and selling the product. Municipalities tend not to be successful at competing with private enterprise. Therefore, in order to utilize this alternative, the grant funding must be sufficient to help make the cost per dry ton competitive with the other processes.

J.   Alternative J - Contract Hauling/Truck/Land Application

The complete sludge handling system for this alternative is:

Belt-pressed   **to**   Contract   **to** Land
Sludge   Hauling   Application

This method of sludge disposal is currently being used by the City of St. Petersburg. It has no technical treatment process or equipment. The dewatered sludge is hauled to a designated site and mechanically applied to the soil. At the extreme, it is landfilled but usually the hauler can find a land owner to purchase it. Currently, most of the City of St. Petersburg's sludge goes to orange groves in Polk Orange, and Lake Counties. The moist sludge is an amendment to the dry sandy soils of the groves. Additionally, the sludge can be used for farming, but the regulations of the DER 17-7 Part IV restricts its use in crops for human consumption. If more disinfection in addition to anaerobic digestion is conducted, the sludge could be more widely used.

Proven reliability and cost-effectiveness of this sludge disposal method has been demonstrated in municipalities throughout Florida. Trucking of dewatered sludge for land application may only become less effective with higher fuel costs and stricter land application rules. As agricultural sites become fewer with new development, the cost of hauling will increase. Therefore, a basic disadvantage of this alternative would be the long-term reliability.

7.3   COSTING FACTORS FOR CAPITAL AND O&M COST ESTIMATES

A.   General

This subsection will provide the listing of costing criteria on which all the tables in Subsection 7.4 (Cost Estimates of Alternatives) are based. The criteria noted here are based on good engineering practice, survey of current conditions, and are acceptable to the City of St. Petersburg staff.

For FSO facilities, the manufacturer of the facility will finance, design, build, operate, maintain, and assume responsibility for the ultimate disposal of the end product. No new additional facilities will be required by the City of St. Petersburg at the wastewater treatment plants. Responsibility for compliance with the State of Florida Regulations regarding sludge will also be met by the full service operation.

B.   Assumptions/Criteria

    1.   Sludge Characteristics -

        a. Percent solids - 15-21% with an average of 17%.

        b. Sludge quantity - as shown in Table 3-6 of this report. Facilities size - 20 dry tons/day, based on a 5-day work week.

        c. With the Industrial Pretreatment Program (IPP), all sludge from the City of St. Petersburg WWTP's should be Grade I (FDER 17-7 Part IV).

        d. No significant change in chemical characteristics of sludge during the planning period. The basic chemical characteristics of an anaerobically digested sludge is:

            Volatile Solids - 45% of Total Solids
            Total Nitrogen - 4.0% of Total Solids
            Total Phosphorous - 3.0% of Total Solids
            Total pathogenic organisms are significantly reduced.

    2.   Annual Escalation Rates -

        a. Power Costs

            (1) Electricity - 7%
            (2) Fuel - 8%

        b. Labor - 6.5%

        c. Spare parts and maintenance - 8%

        d. Chemicals - 8%

        e. General yearly inflation rate - 7%

        f. Escalation allowance prior to bid - 10%

3.   Financial Analysis –

   a. Service life of facility – 20 years
       (except Lime Encapsulation which is 12 years)

   b. Planning Period – 20 years

   c. Discount Factor – 10%

   d. Reinvestment revenue interest – 9%

4.   Unit Costs – 1985

   a. Diesel Fuel – $1.00 per gallon

   b. Electricity – $0.05 per kilowatt-hour

   c. Landfill gas – $0.148 per 1,000 cubic foot (Ref No. 5)

   d. Steam revenue @ 10 psi = 1 kilowatt and $0.042 per
       kilowatt-hour (Ref. No. 5)

   e. Powdered Quicklime (CaO) – $75 per dry ton

   f. Auxiliary Fuel – $1.50 per gallon

   g. Wood Chips – $5.00 per ton (for bulking) and $16 per ton
       (for fuel)

   h. Other Bulking Material – $1.20 per ton

   i. Labor – $15.75 per hour (for most equipment operation).
       Includes overhead.

7.4   COST ESTIMATES OF ALTERNATIVES

A. Alternative A – Rotary Drying/Full Service Operator (Figures 7-2 and
    7-3)

   1.   Summary

      a. Capital Costs

          Total projected costs for this alternative is $7,400,000
          including a 10% escalation.   This cost includes
          additional equipment not included in previous rotary
          drying installations, such as a sewage grinder, a
          hammermill, and a wood chip burning furnace.

      b. Operation and Maintenance Costs

          The major item in O&M Cost is wood chip fuel for the

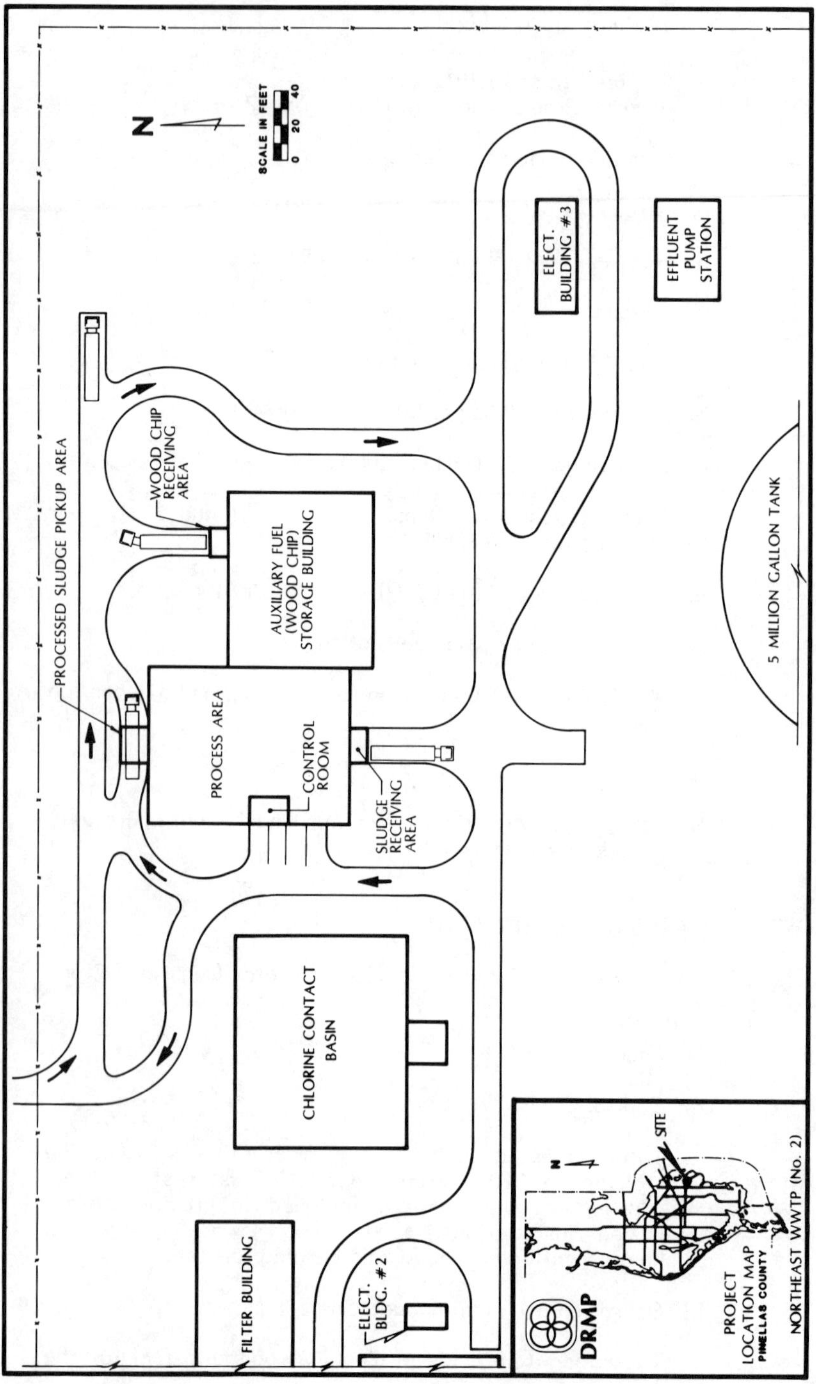

PROPOSED SITE PLAN FOR ROTARY DRYING AT THE N.E. WWTP

Figure 7-2

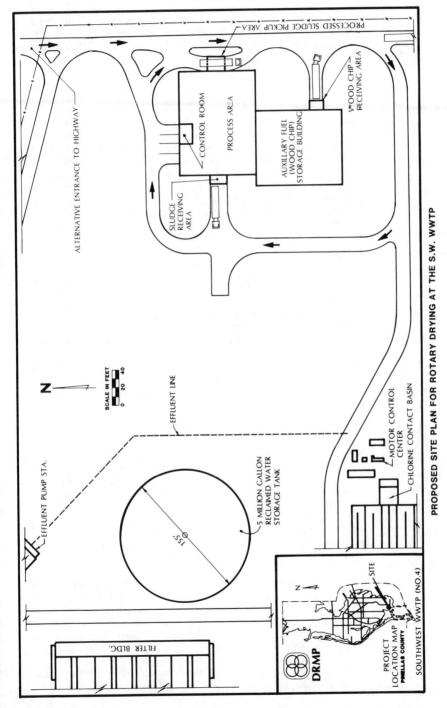

PROPOSED SITE PLAN FOR ROTARY DRYING AT THE S.W. WWTP

Figure 7-3

furnace and the internal hauling costs. The total estimated cost is $585,000. After revenue from the dried product is considered, the cost is $255,000 per year.

2.   Capital Cost

Refering to Table 7-2, the process equipment, conveying systems, and building costs are approximately $4,300,000. This includes the basic drying process plus some optional equipment. The heat for the process is supplied by the wood chip burning furnace. Consequently this operation requires a rather sizable wood chip storage building. The capital cost also includes the cost of a buried diesel fuel tank to provide additional fuel If needed. A hammermill and sewage grinder is used to make the dried product more consistent. Maintaining a consistent product has been a problem in other installations and affects the marketability of the end product.

The manufacturer states that the cyclone separator and the wet scrubber are sufficient to control air pollutants including any extra particulates from the wood burning. In fact, the wood burns cleaner than fuel oil, and it has less sulfur. Two rotary dryers are included, each capable of producing 1,800 pounds per hour of approximately 95% dried sludge. Also included is a dry product silo and conveying system along with a truck scale. The site requirements are shown on the alternative proposed site plans Figure 7-2 and 7-3. The total estimated project cost is $7,400,000 for this alternative.

3.   Operation and Maintenance Costs

Refering to Table 7-3, the basis of the operational and maintenance cost estimates comes from the manufacturer's proposal. Their estimates are used because they are considered accurate and, due to the company's previous operational experiences in Florida, should be appropriate. The labor costs are based on I (one) 2-man crew consisting of a mechanic/foreman and an operator/laborer. It is assumed that these people will also operate the truck scale and perform administrative duties.

Under normal operations with an adequate quantity of wood chips and the furnace in good operating condition, no fuel oil is needed to dry the sludge. Approximately 500 gpm of water is required for the wet scrubbers for which reclaimed effluent water should be used. This amount will then need to be treated along with normal personnel sewer flows.

The rotary dryers require regular lubrication and maintenance, representing the bulk of the equipment

TABLE 7-2

ALTERNATIVE A

ESTIMATED PROJECT COST

ROTARY DRYING/FULL SERVICE OPERATOR

Description	Estimated Cost
1. Site Work/Road	$ 50,000
2. Concrete	300,000
3. Metal Building	800,000
4. Masonry/Metal/Carpentry	10,000
5. Doors/Windows/Glass	10,000
6. Finishes	20,000
7. Process Equipment	3,200,000
8. Conveying Systems	300,000
9. Mechanical	
Piping	40,000
HVAC	80,000
Plumbing	20,000
10. Fire Control	25,000
11. Electrical	250,000
12. Miscellaneous Construction Costs	100,000
Subtotal	$5,205,000
13. Contingency (15%)	781,000
Construction Cost	$5,986,000
14. Engineering & Technical Services (12%)	718,000
Subtotal	$6,704,000
15. Escalation Allowance (10%)	670,000
TOTAL PROJECT COST	**$7,374,000**
TOTAL PROJECT COST (ROUNDED) FY 1987	**$7,400,000**

TABLE 7-3

ALTERNATIVE A

ESTIMATED OPERATION & MAINTENANCE COST

ROTARY DRYING/FULL SERVICE OPERATOR

	Description	Estimated Cost
1.	Labor	$ 90,000
2.	Fuel Oil (Woodchips)	200,000
3.	Water/Sewer (Reclaimed Water)	30,000
4.	Equipment/Building Maintenance	20,000
5.	General Building Utilities	40,000
6.	Chemicals (Carrier Oil)	-0-
7.	Expendables	10,000
8.	Electric Power	30,000
9.	Internal Hauling	120,000

TOTAL OPERATION & MAINTENANCE

COST FY 1987 $ 540,000

| 10. | Marketing and Sales | $ 80,000 |
| 11. | Income from Dried Product | (330,000) |

NET ESTIMATED OPERATIONS COST FY 1987 $ 290,000

maintenance cost. Electric power is for running the forced draft blowers and the turning operation of the rotary drums.

As with all the full service operations, an internal hauling cost is included as well as a marketing and sales expense. The income for the dried product is based on current market values for pelletized dried product. For a conservative estimate, it is assumed the product will be 4.0% nitrogen and command a price of $60.00 per dry ton or a total of $330,000. The net estimated O&M cost for the 1987 would then be $290,000.

B.   Alternative B - Multiple-effect Drying/Full Service Operator (Figures 7-4 and 7-5)

   1.   Summary

       a. Capital Costs

           As presented in Table 7-4 the total capital cost is $6,000,000.

       b. Operation and Maintenance Cost

           The total cost for operation and maintenance as shown in Table 7-5 is $780,000 which includes a cost for marketing the end product. After the revenue of the product is considered at $60.00 per dry ton the cost is reduced to $450,000 for the year 1987.

   2.   Capital Cost

       Refering to Table 7-4, Alternate B Costs, the $3,000,000 for process equipment includes the numerous process units in the Carver-Greenfield Mechanical Vapor Recompression (MVR) system which is then followed by the multiple-effect unit. The MVR system is usually required when drying a sludge that has less than 15% solids concentrations. It concentrates the sludge to 50% to 70% by running off the compressed vapor that it produces. Before this step the wet feed material is "fluidized" in a tank by adding a carrier oil. The sludge is easily pumped even when it is 98% dry solids. The multi-evaporator has two stages which further concentrates the solids by utilizing the vapor heat from the second stage in the first stage. The basic principle behind this reuse of steam is that the first chamber is at a lower pressure so that the evaporation of water can occur even with the lower temperatures induced by the vapor from the second evaporation. The temperature and time in this process, which are equivalent to pasteurization, achieve the desired pathogen destruction.

       Both the waste steam and the sludge must go through a deoiling stage. For the waste steam, oil is recovered with

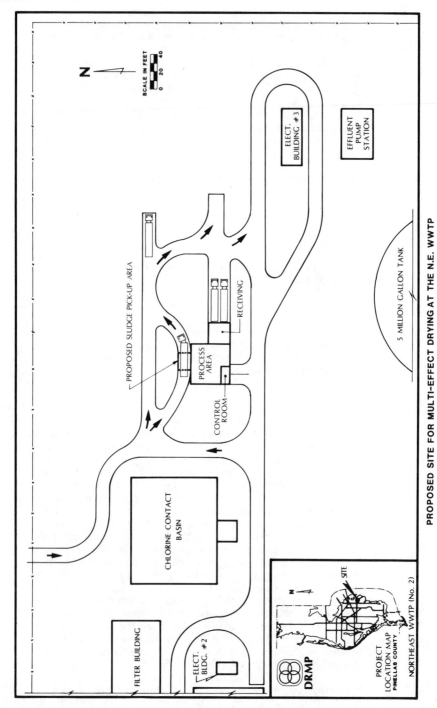

PROPOSED SITE FOR MULTI-EFFECT DRYING AT THE N.E. WWTP

Figure 7-4

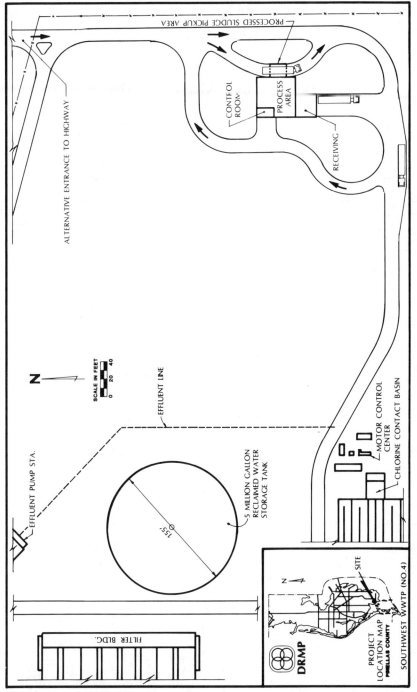

PROPOSED SITE PLAN FOR MULTI-EFFECT DRYING AT THE S.W. WWTP

Figure 7-5

TABLE 7-4

ALTERNATIVE B

ESTIMATED PROJECT COST

MULTIPLE-EFFECT DRYING/FULL SERVICE OPERATOR

Description		Estimated Cost
1.	Site Work/Road	$ 30,000
2.	Concrete	150,000
3.	Metal Building	200,000
4.	Masonry/Metal/Carpentry	10,000
5.	Doors/Windows/Glass	10,000
6.	Finishes	20,000
7.	Process Equipment	3,000,000
8.	Conveying Systems	350,000
9.	Mechanical	
	Piping	50,000
	HVAC	40,000
	Plumbing	20,000
10.	Fire Control	25,000
11.	Electrical	200,000
12.	Miscellaneous Construction Costs	100,000
	Subtotal	$ 4,205,000
13.	Contingency (15%)	630,000
	Construction Cost	$ 4,835,000
14.	Engineering and Technical Services (12%)	580,000
	Subtotal	$ 5,415,000
15.	Escalation Allowance (10%)	540,000
TOTAL PROJECT COST		$ 5,955,000
TOTAL PROJECT COST (ROUNDED) FY 1987		$ 6,000,000

TABLE 7-5

ALTERNATIVE B

ESTIMATED OPERATION & MAINTENANCE COST

MULTIPLE-EFFECT DRYING/FULL SERVICE OPERATOR

	Description	Estimated Cost
1.	Labor	$ 350,000
2.	Fuel Oil	-0-
3.	Water/Sewer	100,000
4.	Equipment/Building Maintenance	30,000
5.	General Building Utilities	30,000
6.	Chemicals (Carrier Oil)	10,000
7.	Expendables	10,000
8.	Electric Power	50,000
9.	IOnternal Hauling	120,000
	TOTAL OPERATION & MAINTENANCE COST FY 1987	$ 700,000
10.	Marketing and Sales Cost	$ 80,000
11.	Income from Dried Product	(330,000)
	NET ESTIMATED OPERATIONS COST FY 1987	$ 450,000

an oil separator, and for sludge a hydroextraction system is required. After the total oil is taken out of the solids an oil recovery system separates the fluidizing oil and the sewage oil. According to the manufacturer, sewage oil is a marketable product. This is an important consideration in that this process, which can handle scum, grease, and septage, can derive a salable product from these items.

Air pollution control devices are minimal, but some pretreatment for odor control should be taken for the vented waste vapor stream. A scrubbing device would be used in a City of St. Petersburg application. The condensate from the process has an odor due to the presence of volatile matter and ammonia. This sidestream is treated and pumped in a closed system to minimize odors. Site requirements are shown on the alternative proposed site plans Figures 7-4 and 7-5.

3.   Operation and Maintenance Costs

It is suggested by the manufacturer that the same person can operate both the belt filter presses and the multiple-effect drying (MED) facility. However, due to distance between the facilities and the level of technology used in the process, it is assumed that a separate person should be assigned to the MED unit. An additional clerical and maintenance personnel to run the facility will be required. The water that is required for process steam and condensing the waste vapor could be obtained from the nearby reclaimed water system. Sewer service is needed for the control building and for the condensate. General building utilities are similar to other alternatives. Power for the start-up and auxiliary heat requirements comes from an electrical source. Internal hauling is required. Marketing and sales cost are added to cover those costs associated with the manpower, advertisement, and possible bagging operation for processed sludge sales. See Table 7-5.

The income for dried product is based upon the current median value for dried sludge as paid by companies currently buying dried sludge from Largo, Florida, Houston, Texas, and Milwaukee, Wisconsin.

C.   Solar Powered Drying/Full Service Operator

I.   Summary

a. Capital costs

The total capital cost for this alternative is $2,900,000 (Table 7-6). Solar Drying is therefore one of the lowest cost alternatives. Due to the limited number of existing installations, the costs are not firm.

TABLE 7-6

ALTERNATIVE C

ESTIMATED PROJECT COST

SOLAR POWERED DRYING/FULL SERVICE OPERATOR [a]

Description		Estimated Cost
1.	Site Work/Road	$ 30,000
2.	Concrete	100,000
3.	Metal Building	150,000
4.	Masonry/Metal/Carpentry	10,000
5.	Doors/Windows/Glass	5,000
6.	Finishes (Arch/Process)	10,000
7.	Process Equipment	1,500,000
8.	Conveying Systems	150,000
9.	Mechanical	
	Piping	20,000
	HVAC	5,000
	Plumbing	20,000
10.	Fire Control	10,000
11.	Electrical	30,000
12.	Miscellaneous Construction Costs	80,000
	Subtotal	$ 2,120,000
13.	Contingency (15%)	320,000
	Construction Cost	$ 2,440,000
14.	Engineering & Technical Services (12%)	290,000
15.	Land	160,000
TOTAL PROJECT COST [b]		**$ 2,890,000**
TOTAL PROJECT COST (ROUNDED) FY 1985		**$ 2,900,000**

(a) Costs listed are not firm due to the newness of the technology.

(b) No escalation will be charged to this alternative; the construction will occur within the next year according to the manufacturer.

b.   Operation and Maintenance Costs

Operation and maintenance costs for this process is mainly for equipment maintenance and electric power for the blowers. The total O&M cost listed in Table 7-7 is $300,000 after revenues from the product sales.

2.   Capital Costs

The capital costs for this process are difficult to clearly define due to the relative newness of the technology. The manufacturer has supplied its estimates, and those costs are used in Table 7-6. The majority of cost is for the solar drying system, pelletizing, and product conveying system. In this alternative, it is assumed that the manufacturer will utilize preowned solar panels to collect the solar heat for drying if this cannot be implemented the cost would increase substantially. The cost is based on the full service operator (Jiffy Industries) supplying the trucks and trailers for the hauling of the dewatered sludge to their facility in Hillsborough County. Therefore, the cost of conveying will include the trucks for sludge hauling. Land cost must be added in to the total project cost. The total cost for the project is then $2,900,000 with numerous caveats.

3.   Operation and Maintenance

Operation and maintenance cost estimates, like capital cost estimates, are primarily based on information supplied by the manufacturer. The major cost items, as shown on Table 7-7, are equipment maintenance and contract hauling cost. Even though the full service operator will provide the truck and trailers, cost for the fuel and labor to haul sludge to Hillsborough County would be incurred by the FSO. Electric power will be required to run the pelletizers and the blower for the process system. Again, some cost for the marketing of the product is included and the income from the product is subtracted from the total O&M cost. The final figure is $300,000 for O & M.

D.   Alternative D - Fluidized Bed Incinerator/Full Service Operator - City Only (Figure 7-6)

1.   Summary

a.   Capital Cost

The capital cost listed on Table 7-8 are $7,300,000 for the City-only fluidized bed incinerator. On the basis of capital costs alone, this alternative is cost-effective in comparison to other high technology treatment processes.

b.   Operation and Maintenance Cost

Operation and maintenance costs of $3,200,000 are very

TABLE 7-7

ALTERNATIVE C

ESTIMATED OPERATION & MAINTENANCE COST

SOLAR POWERED DRYING/FULL SERVICE OPERATOR

	Description	Estimated Cost
1.	Labor	$ 70,000
2.	Fuel Cost (Hauling)	30,000
3.	Water/Sewer	20,000
4.	Equipment/Building Maintenance	200,000
5.	Expendables	30,000
6.	Electric Power	70,000
7.	Hauling (Labor only)	130,000
	TOTAL O & M COST FY 1984	$ **550,000**
8.	Marketing and Sales Cost	80,000
9.	Income from Dried Product	(330,000)
	NET ESTIMATED O & M COST FY 1987	$ **300,000**

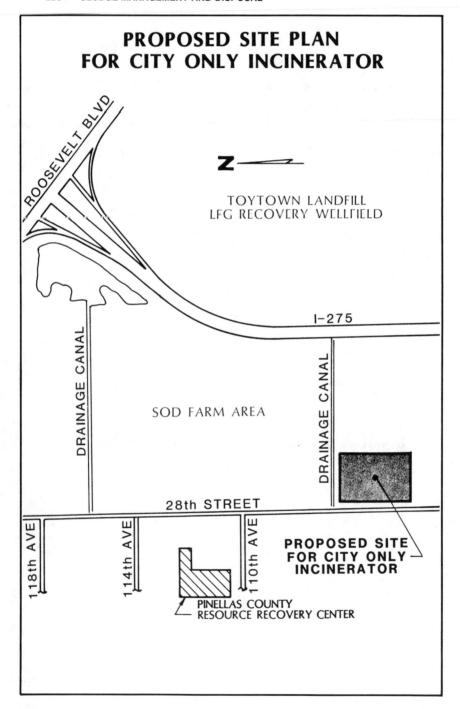

Figure 7-6

TABLE 7-8

ALTERNATIVE D

ESTIMATED PROJECT COST

FLUIDIZED BED INCINERATOR/FULL SERVICE OPERATOR (CITY ONLY)

	Description	Estimated Cost
1.	Site Work/Road	$ 30,000
2.	Concrete	300,000
3.	Metal Building	200,000
4.	Masonry/Metal/Carpentry	30,000
5.	Doors/Windows/Glass	20,000
6.	Finishes (Arch/Process)	20,000
7.	Process Equipment	3,500,000
8.	Conveying Systems	250,000
9.	Mechanical:	
	Piping	30,000
	HVAC	50,000
	Plumbing	10,000
10.	Fire Control	20,000
11.	Electrical	250,000
12.	Miscellaneous Construction Costs	450,000
	Subtotal	$ 5,160,000
13.	Contingency (15%)	770,000
	Construction Cost	$ 5,930,000
14.	Engineering and Technical Services (12%)	710,000
	Subtotal	$ 6,640,000
15.	Escalation Allowance (10%)	660,000
	TOTAL PROJECT COST	**$ 7,300,000**
	TOTAL PROJECT COST (ROUNDED) FY 1987	

high (Table 7-9). They could be mitigated somewhat by either heat recovery, or steam generated electricity revenues. However, the purpose of for this report these items are not considered.

2.   Capital Costs

Table 7-8 lists the individual capital cost items for a City-only fluidized bed incinerator. The process equipment and conveying systems alone will cost $3,750,000. Two 19-foot diameter fluidized bed incinerators will be required to burn the dewatered sludge produced by the City of St. Petersburg.

In the process a blower will be used to fluidize the bed of sand. Also, sludge feeders will discharge the sludge into the hot sand of the incinerator. A wet scrubber, utilizing reclaimed effluent is included. By locating the incinerator at the sod farm, the ash can be pumped into a nearby lagoon. In this situation, additional ash handling equipment is not needed. Also, with the incinerator located on City-owned property, additional costs for land are avoided.

Not included in the capital cost estimates are the equipment for landfill gas utilization and a steam boiler for resource recovery. These items are not considered for this full service operator system serving the City. With the heat produced in the incinerator, steam can be made available for use at the resource recovery plant.

3.   Operation and Maintenance

In Table 7-9, labor costs are similar to any other high technology process, such as multiple-effect drying. The highest costs are for auxiliary fuel and water/sewer. Auxiliary fuel is required to supplement the heat valve of the sludge. An aerobically digested sludge will not sustain combustion alone. If this process is pursued, the digestion times should be reduced in order to have the sludge incineration more autogenous (able to burn without added fuel).

Water will be needed for cooling and for the wet scrubber unit. If reclaimed water is used, some additional treatment will be required. Electric power will be needed for the fluidizing blowers. Each incinerator will have a 400 horsepower blower unit.

The internal hauling cost covers the extra hauling distance to the sod farm area.

TABLE 7-9

ALTERNATIVE D

ESTIMATED OPERATION & MAINTENANCE COST FY 1987

FLUIDIZED BED INCINERATOR/FULL SERVICE OPERATOR (CITY ONLY)

	Description	Estimated Cost
1.	Labor	$ 70,000
2.	Auxiliary Fuel Cost	2,000,000
3.	Water/Sewer	700,000
4.	Equipment/Building Maintenance	100,000
5.	General Building Utilities	20,000
6.	Chemicals	-0-
7.	Expendables	10,000
8.	Electric Power	80,000
9.	Internal Hauling	220,000

TOTAL O & M COST FY 1987 $ 3,200,000

E.   Alternative E - In-Vessel Composting/Full Service Operator (Figures 7-7 and 7-8)

1.   Summary

a.   Capital Cost

The capital cost in this alternative was $8,500,000 (Table 7-10). The major cost items are the process equipment (reactors) and the metal building.

b.   Operation and Maintenance

The major operation and maintenance cost for this process is for carbonaceous material that is used for bulking and adding organics to the sludge for the decomposition process. Total O&M cost is $320,000 including revenue (Table 7-11).

2.   Capital Costs

As with most of the alternatives, the majority of the capital costs for in-vessel composting is the sludge processing equipment. Among the processing equipment is carbonaceous and dewatered sludge storage bins. Carbonaceous material, such as sawdust, bark, newspaper, or wood chips is mixed with the dewatered sludge and then conveyed to the bioreactors. The composting process requires 12-14 days retention time and results in a product that has 60% moisture content and a C:N ratio of 18. This material requires a "post-composting" or curing period which is usually accomplished by a 6-8 week storage period on the ground. In St. Petersburg, however, this is not practical due to the climatic conditions and the possibility of odors and insects. Therefore, a cure reactor will be installed along with a processed sludge storage bin. Note that alternative additional sludge dewatering equipment capable of attaining 35% dry solids with prebulking materials could lower the capital costs by $1.5 million resulting in a cost of $7.0 million. This capital savings is somewhat offset by additional operational costs.

A system of ventilation must be installed to provide oxygen for the composting process. This would replace the plowing over that is provided in a windrow type of composting. Additionally, a rotating discharge device is needed to remove the compost from the bottom of each reactor. The site requirements are shown on the alternative proposed site plans 7-7 and 7-8. Finally, this process will require a CO_2 analyzer and temperature monitoring equipment, as well as other ancillary equipment. These are required to insure that the oxygenation and temperature needs for the process are being met. The total estimated project cost with the existing belt press dewatering opertion is $8,500,000. See Table 7-10.

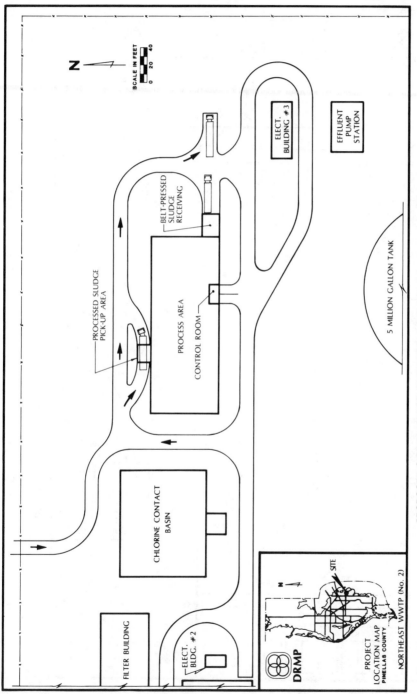

PROPOSED SITE PLAN FOR IN-VESSEL COMPOSTING AT THE N.E. WWTP

Figure 7-7

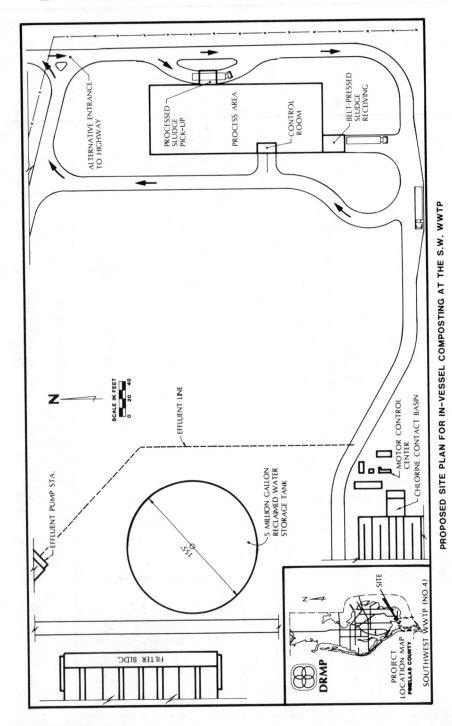

PROPOSED SITE PLAN FOR IN-VESSEL COMPOSTING AT THE S.W. WWTP

Figure 7-8

TABLE 7-10

ALTERNATIVE E

ESTIMATED PROJECT COST

IN-VESSEL COMPOSTING/FULL SERVICE OPERATOR

Description	Estimated Cost
1. Site Work/Road	$ 50,000
2. Concrete	140,000
3. Metal Building	500,000
4. Masonry/Metal/Carpentry	20,000
5. Doors/Windows/Glass	10,000
6. Finishes (Arch/Process)	20,000
7. Process Equipment	4,800,000
8. Conveying Systems	100,000
9. Mechanical	
Piping	20,000
HVAC	30,000
Plumbing	30,000
10. Fire Protection/Water/Sewer	50,000
11. Electrical (non-process)	60,000
12. Miscellaneous Construction Costs	150,000
Subtotal	$ 5,980,000
13. Contingency (15%)	895,000
Construction Cost	$ 6,875,000
14. Engineering and Technical Services (12%)	825,000
Subtotal	$ 7,700,000
15. Escalation Allowance (10%)	770,000
TOTAL PROJECT COST	$ 8,470,000
TOTAL PROJECT COST (ROUNDED) FY 1987	$ 8,500,000

TABLE 7-11

ALTERNATIVE E

ESTIMATED OPERATION & MAINTENANCE COST FY 1987

IN-VESSEL COMPOSTING/FULL SERVICE OPERATOR

	Description	Estimated Cost
1.	Labor	$ 70,000
2.	Power and Fuel	80,000
3.	Water/Sewer	30,000
4.	Equipment/Building Maintenance	50,000
5.	Expendables	20,000
6.	Carbonaceous Material	120,000
7.	Internal Hauling	120,000
	TOTAL O & M COST FY 1987	$ 490,000
8.	Marketing and Sales	$ 60,000
9.	Credit for Compost Product	(230,000)
	NET ESTIMATED OPERATIONS COST FY 1987	$ 320,000

3.   Operation and Maintenance

In Table 7-11 is a description of the various operation and maintenance costs for in-vessel composting. As with rotary drying, the major cost items of the process will be carbonaceous material and internal hauling. Carbonaceous material in this case will be needed for bulking material rather than for heat generation. The biological activity should produce the required heat for this process.

A labor cost is included for the operator on a full time basis due to the many responsibilities, i.e., operation, maintenance, and weighing trucks. Generally, a single operator could perform both tasks of belt-press operation and composting, but due to the physical separation of the processes, that cannot be considered here. Clerical and supervisory type labor for the running of a full service operation is included.

Power and fuel cost is for the operation of the aeration and exhaust blowers and the conveying systems.

The total credit for compost is based on a conservative estimate

of $30/per wet ton of product. The total product generated is 50 wet tons per work day. A marketing and sales cost is also included in the total O&M cost of $320,000.

F.   Alternative F - Lime Encapsulation/Full Service Operator (Figure 7-9)

1.   Summary

a.   Capital Costs

In Table 7-12 are shown the total project costs for the years 1985 and 1997 as $260,000 and $1,300,000, respectively.

b.   Operation and Maintenance Costs

In Table 7-13 is presented the first full year of O&M cost in 1986. The total O&M cost for this alternative is $270,000.

2.   Capital Costs

In analyzing capital costs for the previous alternatives, the service life of the equipment was 20 years. In this process the service life is between 10 to 14 years or an average of 12 years. Therefore, the capital cost for the years 1985 and 1997 are estimated in Table 7-12.

Each of the four plants will be equipped with one (1) Roediger Lime Post Treatment (LPT) unit RM 240 with a nominal capacity

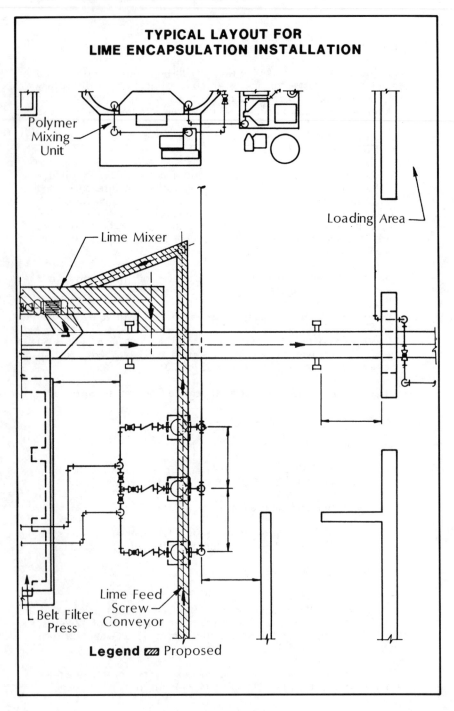

**TYPICAL LAYOUT FOR
LIME ENCAPSULATION INSTALLATION**

Polymer Mixing Unit

Loading Area

Lime Mixer

Belt Filter Press

Lime Feed Screw Conveyor

Legend ▨ Proposed

Figure 7-9

TABLE 7-12

ALTERNATIVE F

ESTIMATED PROJECT COST

LIME ENCAPSULATION/FULL SERVICE OPERATOR

Description	Estimated Cost
Equipment & Installation[a]	$ 505,000
Storage Facilities	200,000
Sitework 55,000	
Contained and Product Handling Facilities	220,000
Conveying System	160,000
Subtotal $1,140,000	
Contingency (15% of equipment & installation)	171,000
Construction Cost	$ 1,311,000
Engineering & Technical Service (12%)	157,000
TOTAL PROJECT COST	$ 1,468,000
x45% (Local Share)[b]	$ 661,000
TOTAL ROUNDED LOCAL SHARE FY 1985	$ 660,000

Notes:

(a) LPT mixer unit with conveyer, silo, and installation.

(b) Assuming 55% FDER Wastewater Construction Grant Funding.

TABLE 7-13

ALTERNATIVE F

ESTIMATED OPERATION & MAINTENANCE COST

LIME ENCAPSULATION/FULL SERVICE OPERATOR

Description	Estimated Cost
Labor (one operator per WWTP)	$ 140,000
Chemical (Quicklime)	108,000
Electricity	3,000
TOTAL	$ 251,000

NET OPERATION & MAINTENANCE COST FY
(ROUNDED) 1986 $ 250,000

of 2000 dry pounds of solids per hour. The Northwest Plant #3 and Southwest Plant #4 each will have one additional LPT unit as backup. The capital cost also includes the costs for covered storage area, associated sitework, new conveying system, and storage containers.

The total project cost is $1,468,000 of which the City pays 45% or $660,000 due to expected agency construction grant participation.

In 1997, new process equipment will be purchased. The total project cost based on equipment and 3% for engineering is estimated at $1,170.00. The City will be responsible for 100% of the purchase and installation costs.

3.   Operation and Maintenance

For a conservative estimate, the assumption was made that a separate operator would be assigned to the Lime Post Treatment (LPT) unit at each plant. Belt filter press operation is designed for an 8-hour day. The Roediger unit operates in conjunction with the belt presses. Therefore, the FSO operation is 40 hours per week.

The quicklime (CaO) is added at the rate of 0.25 dry tons of CaO per dry ton of sludge. The 1985 base cost of CaO is $75 per dry ton.

Table 7-13 presents the costs of O&M without considering (1) revenues from the material, or (2) cost for hauling. It is assumed that the product is available for pick-up without cost to the City or the hauler.

G.   Alternative G - City/County Incineration Project
(For site plan see Pinellas County Regional Sludge/Grease Disposal Facility in Appendix C, Attachment 5)

1.   Summary

Under this alternative is the plan for Pinellas County to build a sludge incinerator to serve the entire county. The incinerator will also burn grease and landfill gas. As the largest city in Pinellas County, the City of St. Petersburg would be the major user. The possibility of grant funding from the Department of Environmental Regulations helps make this alternative worthwhile for review.

Table 7-14 lists information on the cost of the proposed incinerator. No itemization is available on the capital or operation and maintenance costs. The capital cost is listed as $17,725,000 and yearly O & M cost as $3,098,800.

TABLE 7-14

ALTERNATIVE G

SUMMARY OF COSTS

CITY/COUNTY INCINERATION PROJECT[a]

Description	Estimated Cost[a]
Capital Costs	$ 17,725,000[b]
Engineering @ 12%[c]	$ 2,127,000
Total Project Cost	**$ 19,852,000**
Yearly O & M Costs	$ 3,098,800
Yearly Revenue	$ 1,304,000
Average Annual Costs	$ 3,650,700
Average User Fee/Wet Ton for 1987[d]	**$ 17.45**
Average User Fee/Dry Ton for 1987[d]	**$ 102.65**
Average User Fee/Dry Ton Rounded for 1987[d]	**$ 105.00**

NOTES:

(a)   From County of January 28, 1985.

(b)   Based on 3-24 ft. diameter incinerators installed ultimately.

(c)   Estimated.

(d)   Annual costs and user costs based on phased construction of incinerator units. Assuming a 14 million dollar capital cost.

(e) $25 per Dry Ton added for internal hauling costs.

2.   Capital Cost

The proposed facilities for the Pinellas County Regional Sludge/Grease Disposal Facility consists basically of:

- ultimately 3-24 foot diameter fluidized bed incinerators.
- a grease handling and pretreatment facility
- a landfill gas recovery system.
- a steam driven turbine generator

The site is located just north of the Pinellas County Resource Recovery Facility.

3.   Operation and Maintenance

As discussed under Section 7.4, D, for the City-only incinerator, the operation of a fluidized bed incinerator is labor intensive. It also requires considerable fuel to burn sludge. The utilization of landfill gas and grease in this alternative reduces the requirement for auxiliary fuel. Another considerable O & M cost is the water for cooling towers and air scrubbers. The source of the water would be treated wastewater effluent. Spent water would be treated in the wastewater treatment plant.

H.   Alternative H - Lime Encapsulation/City-Owned (Figure 7-6)

The capital costs for this alternative is found in Table 7-12 Alternative F-Lime Encapsulation. Alternatives F and H have the same capital costs, but differing approaches to operation. Alternative H allows for a City owned and operated facility.

The operation and maintenance of the equipment is relatively uncomplicated. This will allow the belt press operator to spend half his time operating and maintaining the lime encapsulation process.

Therefore, the 1986 O & M cost for this alternative is $180,000 as presented on Table 7-15.

I.   Alternative I - In-Vessel Composting/City-Contract

In this alternative, the City of St. Petersburg would purchase the facility. The operations would be contracted out to the full service operator. This alternative is included to examine how possible Federal and/or State funding for in-vessel composting would affect tipping fees. The capital and O & M costs are the same as shown on Table 7-10 and 7-11.

J.   Alternative J - Truck/Land Application

TABLE 7-15

ALTERNATIVE H

ESTIMATED OPERATION & MAINTENANCE COST

LIME ENCAPSULATION/CITY OWNED AND OPERATED

Description	Estimated Cost
Labor (one operator half-time per WWTP)	$ 70,000
Chemical (Quicklime)	108,000
Electricity	3,000
TOTAL	$ 181,000
NET OPERATION & MAINTENANCE COST FY (ROUNDED) 1986	$ 180,000

No facilities, equipment, or land is required by the City for this alternative due to the contractual arrangement. The successful bidder will supply all the necessary equipment to collect the dewatered sludge at each of the four wastewater treatment plants and subsequently haul and apply it to the site that has been retained by the hauler. Actual hauling costs will be used to project tipping fees.

7.5   SURVEY OF TECHNICAL PROCESSES USED IN THE TEN ALTERNATIVES FOR SLUDGE MANAGEMENT

Users of each technical process on which the ten favorable alternatives are based, were identified and surveyed in order to answer a set of questions about each technology.   Those processes surveyed include solar drying, multiple-effect drying, rotary drying, lime encapsulation, in-vessel composting, and fluidized bed incineration.

Questions asked of individuals familiar with one of the above technologies included the size of installation, types of wastes producing the sludge, costs of the installation, operation and maintenance costs, and any maintenance problems experienced. Results of this survey are to be found in Exhibit 7-1 through 7-6. These exhibits were the most indicative of the technology based upon the survey responses.

EXHIBIT 7-1

SLUDGE TREATMENT INSTALLATION SURVEY

1.   Technology Used: **Rotary Drying**

2.   Location: **Largo, Florida**

3.   Contact: **Eric Blankman (Largo WWTP Operator)**

4.   Year Built: **1976**

5.   Capital Cost: **$900,000**

6.   Operation and Maintenance Costs:

> **No idea - no metering devices built into system.   Contract mentioned process was energy intensive.   Approximately, $200 per dry ton.**

7.   Sludge from what type of waste?

> **Mostly domestic wastes. Less than 5% industrial wastes.**

8.   Sludge Handling Capacity.

> **2.5-3 tons dry solids/day**

9.   Final Product:

> a.   % Solids: **97%**
>
> b.   Sold as what?   **Commercial Fertilizer**
>
> c.   For how much? **$88/dry ton**
>
> d.   Other

10.   Operation problems experienced?

> **Only ordinary, normal maintenance. Handles abrasive water.**

> *   Note that numerous complaints of odors have been filed against this drying installation.

11.   Responsiveness of the manufacturer for service.

> **Had to rebuild air handling system twice in the last 9 years.**

EXHIBIT 7-2

SLUDGE TREATMENT INSTALLATION SURVEY

1. Technology Used: **Multiple-Effect Drying (Carver Greenfield Process)**

2. Location: **Coors Brewery - Golden, Colorado**

3. Contact: **Bart Lynam (Manufacturer's Representative)**

4. Year Built: **1977**

5. Capital Cost: **$3.5 Million**

6. Operation and Maintenance Costs:

   $60-80/dry ton

7. Sludge from what type of waste?

   Brewery Waste

8. Sludge Handling Capacity.

   30 tons dry solids/day

9. Final Product:

   a. % Solids: **68%**

   b. Sold as what? **Pelletized taken to landfill, soon to be sold as animal feed**

   c. For how much? N/A

   d. Other N/A

10. Operation problems experienced?

    No, because process is simple.

11. Responsiveness of the manufacturer for service.

    There has been no unscheduled downtime due to mechanical failures.

EXHIBIT 7-3

SLUDGE TREATMENT INSTALLATION SURVEY

1. Technology Used: **Solar Drying (Jiffy Industries)**

2. Location: **Sorrento, Florida**

3. Contact: **Frasier Bingham (Jiffy Industries)**

4. Year Built: **Not Complete**

5. Capital Cost: **Between $1.7 to $2.0 Million. Does not include land costs.**
       10 acres required

6. Operation and Maintenance Costs:

       $43/dry ton (includes hauling costs)

7. Sludge from what type of waste?

       Municipal

8. Sludge Handling Capacity.

    20 tons dry solids/day

9. Final Product:

    a.   % Solids: **95%**

    b.   Sold as what?   **Fertilizer**

    c.   For how much? **Not Available**

    d.   Other

10. Operation problems experienced?

       Not Available

11. Responsiveness of the manufacturer for service.

       Not Available

EXHIBIT 7-4

SLUDGE TREATMENT INSTALLATION SURVEY

1. Technology Used: **Fluidized Bed Incinerator Flow Solids Reactor with Hot Wind Box**

2. Location: **Kansas City, Kansas**

3. Contact: **Myron Cailterux (Kansas City Municipal Employee)**

4. Year Built: **1979**

5. Capital Cost: **$2.0 Million ($50,000 for waste heat boiler)**

6. Operation and Maintenance Costs:

    $300/dry ton. Costs are lower when incineration receives higher loadings.

7. Sludge from what type of waste?

    Municipal Waste (primary sludge)

8. Sludge Handling Capacity.

    8-9 tons dry solids/day operating at 1/2 capacity

9. Final Product:

    a.   % Solids: Ash

    b.   Sold as what?   Disposed

    c.   For how much?   Not Applicable

    d.   Other   **Inert ash - hauled away.**

10. Operation problems experienced?

    Sludge conditioning equipment gets clogged, had to retrofit boiler.

11. Responsiveness of the manufacturer for service.

    Provide own service.

EXHIBIT 7-5

SLUDGE TREATMENT INSTALLATION SURVEY

1. Technology Used: In-Vessel Composting

2. Location: Portland, Oregon

3. Contact: Ron Dye (Manufacturer Representative)

4. Year Built: May, 1984 (System Not Yet Fully Operational)

5. Capital Cost: $12.0 Million

6. Operation and Maintenance Costs:

    $18.50/dry ton

7. Sludge from what type of waste?

    Municipal Wastes

8. Sludge Handling Capacity.

    60 tons dry solids/day

9. Final Product:

    a.   % Solids: 55%

    b.   Sold as what?   Garden Grow Fertilizer

    c.   For how much? $10–40/ton

    d.   Other

10. Operation problems experienced?

    Still working out problems with conveying systems.

11. Responsiveness of the manufacturer for service.

    Replacement mechanical equipment can be purchased in this country except the rotating screw which is manufactured in Germany.

EXHIBIT 7-6

SLUDGE TREATMENT INSTALLATION SURVEY

1. Technology Used: **Lime Encapsulation**

2. Location: **Pittsburgh, Pennsylvania**

3. Contact: **Mike Slamang (Pittsburgh Municipal Employee)**

4. Year Built: **Currently in design stages for full scale. A pilot scale has been**
    previously tested.

5. Capital Cost: **$100,000**

6. Operation and Maintenance Costs:

    $65/ton for lime.
    10-15% CaO dry ton/dry solid.

    Maintenance is minimal - process is simple.

7. Sludge from what type of waste?

    Will handle anything.

8. Sludge Handling Capacity.

    Any amount tons dry solids/day

9. Final Product:

    a.   % Solids: **30-35%**

    b.   Sold as what?   **Fertilizer**

    c.   For how much? **Not Available**

    d.   Other

10. Operation problems experienced?

    None - process going from experimental to design stage.

11. Responsiveness of the manufacturer for service.

    Can buy equipment anywhere. Simple Equipment.

SECTION 8

COMPARATIVE ANALYSIS OF ALTERNATIVES UTILIZING TRANSFORMATION CURVES

8.1   GENERAL

This section compares the top ten alternatives presented in Section 7. The comparative evaluation is accomplished by Transformation Curve Analysis, as described in Subsection 8.2. The results of the analysis are described in Subsection 8.4.

The most favorable alternatives determined from this analysis are further investigated in Section 9.

8.2   TRANSFORMATION CURVE ANALYSIS

The objective of this sludge disposal investigation is to present the City with a list of alternatives which can effectively dispose of sludge. Preliminary estimates of the capital and annual costs are provided in Section 7 and lists of the associated advantages and disadvantages of each process technology are discussed in Section 4 and 5.

The analytical technique involving the use of transformation curves provides for the analysis of numerous alternatives to arrive at a set of more favorable alternatives. A transformation curve provides an illustrative means in which multiple objective trade-offs can be displayed. The transformation curve (as used in this report) is that curve which describes those comparatively most favorable alternatives with respect to delineated objectives. The construction of the curve is determined by (1) the envelope of the extreme alternative points with respect to the objectives, and (2) by the restriction that the slope of all tangents to the curve must be negative.

The objectives identified to be incorporated in this analysis are the (1) estimated capital cost, (2) estimated annual cost, and (3) evaluation criteria indexes. A reciprocal cost relationship provides an abscissa or ordinate with increasing favorability with distance from the origin. Therefore, the relationship for the estimated capital cost was chosen as $10,000,000 divided by the capital cost estimate.

Likewise, the relationship for the annual cost was chosen as $1,000,000 divided by the annual cost estimate. The resulting ratios are plotted in a field.

The alternatives which comprise the transformation curve are the most favorable with respect to the objectives considered. Alternatives which do not lie on the transformation curve are comparatively less favorable. The trade-off between each objective considered in the analysis is relative to the slope of the tangent at that point.

8.3   EVALUATION FACTORS/CONSIDERATIONS

A.   Evaluation Criteria

In Table 8-1 is a list of the ten criteria selected for the evaluation process which are discussed in Subsection 8.3. The ten criteria are considered the most important by the City of St. Petersburg staff and the engineering project group. Each criterion is given a weighting

TABLE 8-1

SELECTION CRITERIA CATEGORIES[a]

Operation and Maintenance Considerations

- Reliability
- Ease of Operation and Maintenance
- Sidestream Treatment
- Fuel and Chemical Requirements

Community Welfare Considerations

- Public Health
- Community Acceptance
- Environmental Impact

Other Engineering Considerations

- Successful Experience
- Marketability of end-Product
- Implementability

NOTES:

(a)   Each criteria is on a scale of 1 to 10, where 10 is the best score.

factor of 1-10. A zero (0) value is cause for rejection, while a value of ten (10) indicates the alternative is the best in the criteria.

The various criteria are categorized into a general heading of:

-    Operation and Maintenance Considerations,
-    Welfare of the Community,
or   -   Other Engineering Considerations.

A total index value for an alternative in each category is comprised of the scores for the individual criteria in that category. The scores are then used in the plotting of transformation curves.

Through a comparison based on numerous factors other than just costs, the most favorable alternatives can be better selected. This analysis considers the ten criteria listed in Table 8-1 as each relates to cost and quantifies that relationship by the criteria evaluation index. In this way, every alternative is subjected to the same list of criteria. The results of the comparison are placed in tables (Tables 8-2 through 8-5) where they can be readily surveyed. In the second part of the sensitivity analysis, transformation curves are drawn from these relationships between costs and evaluation criteria index as is discussed below.

B.   Operation and Maintenance Considerations

1.   Reliability

This significant evaluation factor is made of many items. In assigning a value to this criterion, each alternative was reviewed for such things as:

-    Mechanical downtime, either for regular maintenance or unscheduled.

-    Amount of storage for essential chemicals or materials.

-    Duplication of equipment or back-up equipment availability.

-    Reliability of management approach or technique employed in the alternative.

Some consideration is given to experience with reliability for the technology, or conversely the lack of any experience.

2.   Ease of Operation and Maintenance

This criterion compares the alternatives according to ease of running and maintaining the process. The level of difficulty can be gauged by either degree of operator skill required, the number of instruments controlling the process, or the amount of training

TABLE 8-2

OPERATION AND MAINTENANCE CRITERIA
COMPARATIVE EVALUATION
(A Total of 40 Points Possible)

Alternative	Description	Evaluation Criteria					Total Index Value
		Reliability	Ease of O & M	Sidestream Treatment	Fuel and Chemical		
A	Rotary Drying/F.S.O.	7	6	8	6		27
B	Multiple-effect Drying/F.S.O.	10	6	5	8		29
C	Solar Drying/F.S.O.	2	6	6	8		22
D	Fluidized Bed Incineration/F.S.O.	8	4	6	4		22
E	In-Vessel Composting/F.S.O.	7	8	10	6		31
F	Lime Encapsulation/F.S.O.	4	9	10	4		27
G	F.B. Incineration/City-County	7	4	8	6		25
H	Lime Encapsulation/City	8	8	10	4		30
I	In-Vessel Composting/City	7	8	10	6		31
J	Contract Hauling/Truck	10	8	10	6		34

TABLE 8-5

SUMMARY COMPARATIVE EVALUATION
AND COSTS FOR THE SELECTED ALTERNATIVES

Alt.	Description	Capital Costs (a)	Operation & Maintenance Cost (a)	Evaluation Criteria		
				Operation & Maintenance Considerations	Welfare of the Community	Other Engineering Considerations
A	Rotary Drying/F.S.O.	$ 7,400,000	$ 290,000	27	19	23
B	Multiple-effect Drying/F.S.O.	6,000,000	450,000	29	28	26
C	Solar Drying/F.S.O.	2,900,000	300,000	22	21	18
D	Fluidized Bed (F.B.) Incineration/F.S.O.	7,300,000	3,200,000	22	22	24
E	In-Vessel Composting/F.S.O.	8,500,000	320,000	31	28	22
F	Lime Encapsulation/F.S.O.	660,000(b)	250,000	27	22	19
G	F.B. Incineration/City–County(c)	—	1,180,000(d)	25	25	26
H	Lime Encapsulation/City(b)	660,000	180,000	30	23	23
I	In-Vessel Composting/City	8,500,000	320,000	31	28	21
J	Contract Hauling/Truck(c)	—	462,000(e)	34	18	29

NOTES:

(a) From Section 7 Tables. All costs are for 1987.
(b) 1985 Costs.
(c) Only annual user costs for this alternative. A one (1) has been assumed for the capital costs.
(d) Based on average user cost in Table 9-3 and 5,372 dry tons.
(e) Based on 15% solids, 110 cubic yards/day, $11.50/cubic yard, from Table 9-6.

OTHER ENGINEERING CONSIDERATIONS

COMPARATIVE EVALUATION

(A Total of 30 Points Possible)

Alternative	Description	Evaluation Criteria			Total Index Value
		Successful Experience	Marketability of End-Product	Implementability	
A	Rotary Drying/F.S.O.	6	9	8	23
B	Multiple-effect Drying/F.S.O.	9	9	8	26
C	Solar Drying/F.S.O.	2	9	7	18
D	Fluidized Bed (F.B.) Incineration/F.S.O.	8	9	7	24
E	In-Vessel Composting/F.S.O.	6	8	8	22
F	Lime Encapsulation/F.S.O.	4	7	8	19
G	F.B. Incineration/City-County	8	9	9	26
H	Lime Encapsulation/City	7	6	10	23
I	In-Vessel Composting/City	6	7	8	21
J	Contract Hauling/Truck	10	9	10	29

TABLE 8-5

SUMMARY COMPARATIVE EVALUATION AND COSTS FOR THE SELECTED ALTERNATIVES

Alternative	Description	Capital Costs (a)	Operation & Maintenance Cost(a)	Evaluation Criteria			
				Operation & Maintenance Considerations	Welfare of the Community	Other Engineering Considerations	
A	Rotary Drying/F.S.O.	$ 7,400,000	$ 290,000	27	19	23	
B	Multiple-effect Drying/F.S.O.	6,000,000	450,000	29	28	26	
C	Solar Drying/F.S.O.	2,900,000	300,000	22	21	18	
D	Fluidized Bed (F.B.) Incineration/F.S.O.	7,300,000	3,200,000	22	22	24	
E	In-Vessel Composting/F.S.O.	8,500,000	320,000	31	28	22	
F	Lime Encapsulation/F.S.O.	660,000(b)	250,000	27	22	19	
G	F.B. Incineration/City-County(c)	1	1,180,000(d)	25	25	26	
H	Lime Encapsulation/City(b)	660,000	180,000	30	23	23	
I	In-Vessel Composting/City	8,500,000	320,000	31	28	21	
J	Contract Hauling/Truck(c)	1	462,000(e)	34	18	29	

NOTES:

(a) From Section 7 Tables. All costs are for 1987.
(b) 1985 Costs.
(c) Only annual user costs for this alternative. A one (1) has been assumed for the capital costs.
(d) Based on average user cost in Table 9-3 and 5,372 dry tons.
(e) Based on 15% solids, 110 cubic yards/day, $11.50/cubic yard, from Table 9-6.

needed to operate and maintain the system.

Also considered here is how difficult the maintenance activities are. This is, for instance, whether the operator can accomplish the task or would the manufacturer representative be required.

3.   Sidestream Treatment

For this criterion, each alternative is reviewed for the amounts of by products and sidestreams that require treatment and the difficulty of such treatment. The additional space for storage, handling, or treatment facilities is considered.   Also, the additional labor, permitting, and impact on the wastewater treatment plant receiving the sidestream is examined.

4.   Fuel and Chemical Requirements

Added fuel, bulking agents, or chemicals for a process cause increased handling and storage requirements. When fuel is stored on site there are safety risks and possible spillage. All these items are considered under this criterion.

C.   Welfare of the Community

1.   Public Health

Each alternative is reviewed for how successful the alternative is in protecting the public health. The following concerns are reviewed:

-    In transporting the end product or dewatered sludge to the treatment facility, what is the exposure to pathogenic organisms?

-    In land application or disposal of end product, what is the likelihood of metal or viral contamination?

-    In the sludge treatment process, what increase, if any, is there in air pollution?

2.   Community Acceptance

Under this criterion, each alternative is evaluated for the degree of acceptance by the community before and after the facility is built. The sludge treatment technology will be reviewed on how well odors and noise can be controlled. Also, the technology is reviewed for success of any previous public awareness efforts conducted.

3.   Environmental Impact

Under this criterion, the technology utilized and the ultimate disposal method employed for each alternative is reviewed. The degree of environmental impact is determined by whether the process technology or method of ultimate disposal will degrade the environment in either land, water, or air quality. They are also evaluated on how well they prevent heavy metals that may be present in the sludge from impacting the environment.

D.   Other Engineering Considerations

1.   Successful Experiences

In this criterion, each alternative is evaluated on the number of installations and the success rate for those installations. Each technology used in an alternative, was researched for the number of installations as listed on Exhibits 4-1 through 4-13. Among these installations, those in Florida and those most closely resembling the proposed installation for the City of St. Petersburg, were reviewed. Low scores were given to those alternatives with unproven technologies. Although this does not reflect upon their reliability, it is difficult to prove reliability without successful installations or pilot plant studies.

2.   Marketability of End Product

This criterion is most important for the alternatives that rely upon a market value of an end product to subsidize the capital debt for a sludge processing facility. In some degree, this is important for all of the alternatives.

Alternatives that used very high capital intensive technologies, such as fluidized bed incineration and in-vessel composting, need a substantial revenue from waste heat or compost, respectively, to make the process cost-effective. Some technologies for sludge treatment, such as lime encapsulation, do not require revenues to be cost competitive with other technologies.

End products have various markets and command differing prices. Some markets are well established such as for dried sludge. Other markets are not established, but have a promising future such as composting. Each alternative, therefore, is reviewed on the circumstances surrounding their end product's marketability.

3.   Implementability

Implementation of an alternative includes many activities. The availability of the resources to use the process is essential. These resources are items such as carbonaceous material for bulking in in-vessel composting, powdered quicklime for lime encapsulation, and landfill gas for a fuel source in the fluidized bed incinerator.

Time requirement for design, construction, and start-up is also a consideration.   Some   processes   require   only   minimal implementation time (lime encapsulation) and some require two or more years (fluidized bed incineration).

8.4   COMPARATIVE EVALUATIONS

A.   General

Presented in Tables 8-2 through 8-5 are the alternatives and the numerical values assigned for each criteria, and Figure 8-1 through 8-7 represent the graphical transformation curve analysis.

The comparative evaluations developed through the curves point out the alternatives which meet the evaluation criteria most cost-effectively.

Alternatives G and J are included in this evaluation. Their capital costs are assumed as one rather than zero so they could be handled mathematically. The annual cost listed is the total yearly user cost for the particular alternative.

B.   Transformation Curve Results

Table 8-6 presents a summary of results for the transformation curve analysis. Listed under the transformation curve number is an "X" indicating the alternatives that are the points that form the envelope of extremes or they lie to the right of that curve.

As a cut-off point, those alternatives that lie on the transformation curves at least one (1) time are considered for further comparison in Section 9. Table 8-7 presents the most favorable alternatives resulting from the transformation curve analysis.

8.5   SUMMARY OF COMPARATIVELY FAVORABLE ALTERNATIVES

The most favorable alternatives, as determined by the investigations and analyses conducted in this section include:

1.   Alternative J is contract hauling and disposal using trucks for transportation. This is shown as the best alternative presently;

2.   Alternative H consists of the conventional city owner and operator method utilizing the lime encapsulation technique. This alternative is shown worthy of experimentation;

3.   Alternative E, in-vessel composting, also utilizes the full service operator technique for procurement, design, construction, financing, permitting, operations, marketing and final disposal of the sludge derived product. This alternative with alternative I were the best capital facility alternatives;

4.   Alternative I incorporates a blend of the conventional City owner with a contracted full service turnkey design, build, operate, market, permit and disposal of the sludge product from an in-vessel composting facility; and

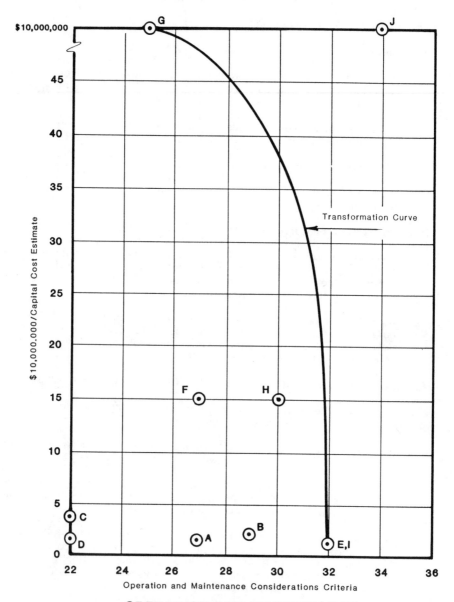

**OPERATION AND MAINTENANCE
VERSUS CAPITAL COST**

Figure 8-1

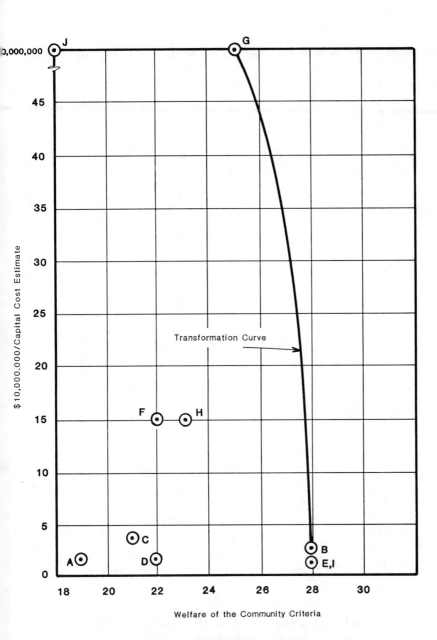

WELFARE OF THE COMMUNITY EVALUATION VERSUS CAPITAL COST

Figure 8-2

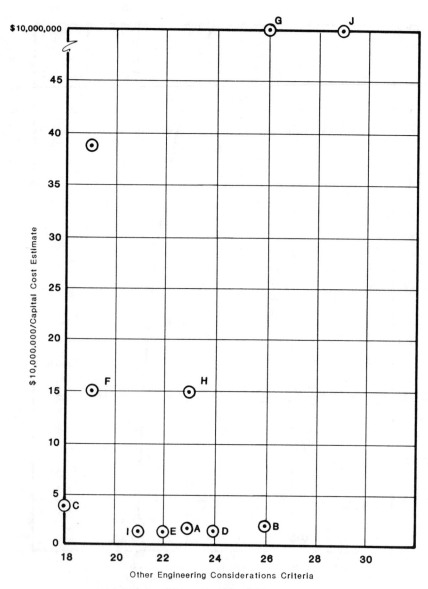

OTHER ENGINEERING CONSIDERATIONS
VERSUS CAPITAL COST

Figure 8-3

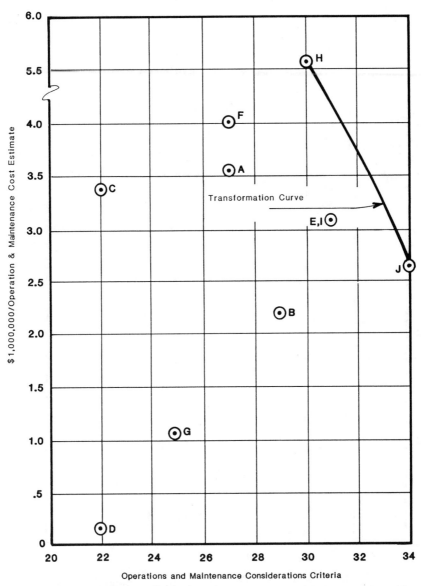

**OPERATION AND MAINTENANCE CONSIDERATIONS
VERSUS OPERATION AND MAINTENANCE COST**

Figure 8-4

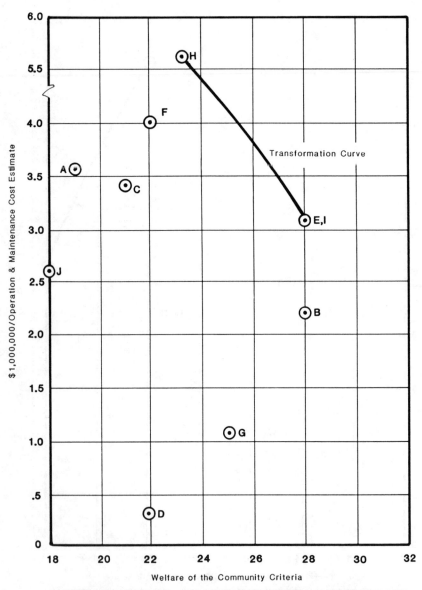

WELFARE OF THE COMMUNITY VERSUS OPERATION AND MAINTENANCE COST

Figure 8-5

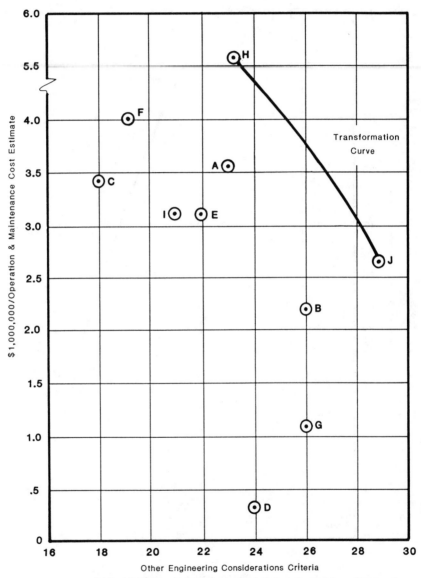

OTHER ENGINEERING CONSIDERATIONS VERSUS
OPERATION AND MAINTENANCE COST

Figure 8-6

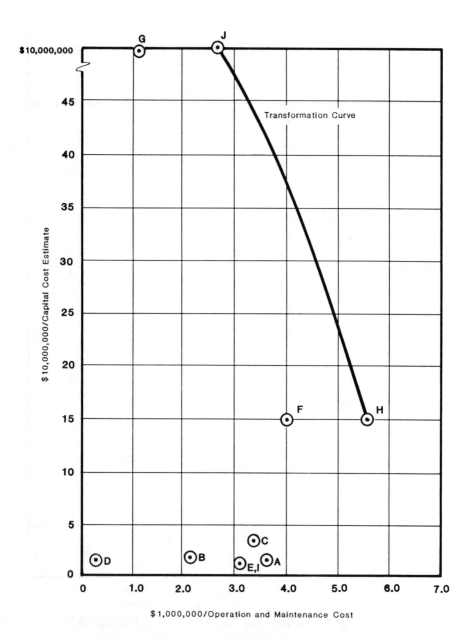

**OPERATION AND MAINTENANCE COST VERSUS
CAPITAL COST**

Figure 8-7

TABLE 8-6

RESULTS

OF THE TRANSFORMATION CURVES

Alter-native	Transformation Curve							
	(1)	(2)	(3)	(4)	(5)	(6)	(7)	Total
A								0
B		X						1
C								0
D								0
E	X	X			X			3
F								0
G	X	X						2
H				X	X	X	X	4
I	X	X			X			3
J	X		X	X		X	X	5

TABLE 8-7

THE MOST FAVORABLE SLUDGE DISPOSAL ALTERNATIVES

Alternative	Technique	Technology
J	Contract Hauling	Truck, Landspreading
H	City	Lime Encapsulation, Truck, Landspreading
E	Full Service Operator	In-vessel Composting/D & M
I	City/Contract	In-Vessel Composting/D&M
G	City/County	Incineration/Resource Recovery
B	Full Service Operator	Multiple-effect Drying/ D & M[a]

Notes:

(a) D & M - Distribution and Marketing

5. Alternative G, which is the City/County incinerator project, provides resource recovery and the ability to incorporate federal and state agency participation in the construction grants program. The City would have no equity invested in the project; and

6. Alternative B, multiple-effect drying utilizing the full service operator technique for procurement of design, construction, financing, operations, permitting, marketing and final disposal of the sludge product.

Alternative J -contract hauling appeared in the favorable region five (5) times out of seven; more than any other alternative. The six alternatives in Table 8-7 are therefore worthy of further consideration in Section 9. The cost analysis in Section 9 and 10 will decide the recommended alternative among the six.

SECTION 9

USER FEES AND PAYOUT PERIOD ANALYSES
OF THE FAVORABLE ALTERNATIVES

9.1   GENERAL

In this section of the report the six (6) alternatives selected as most favorable are compared by two costing methods. The first comparison is "user fee analysis." The cost of treatment per dry ton of sludge is calculated for each alternative and compared.

The second method used for weighing the alternatives is by "pay-out period." This is done by calculating the amount of time to recover the capital cost in revenues or reduced O & M costs.

The most favorable alternatives, as determined by the investigations and analyses conducted for this report, include:

1.   Alternative B - multiple-effect drying utilizing the full service operator technique for procurement of design, construction, financing, operations, permitting, marketing and final disposal of the sludge product.

2.   Alternative E - in-vessel composting also utilizing the full service operator technique for procurement, design, construction, financing, permitting, operations, marketing and final disposal of the sludge derived product.

3.   Alternative G - City/County incinerator project, provides resource recovery and the ability to incorporate Federal and State agency participation in the construction grants program.

4.   Alternative H - City owned and operated method utilizing the lime encapsulation technique.

5.   Alternative I - in-vessel composting incorporating a blend of the conventional City owner with a contracted full service turnkey design, build, operate, market, permit and disposal of the sludge product from an in-vessel composting facility (This approach can utilize the opportunity of construction grant funding while increasing the City's assets.)

6.   Alternative J - contract hauling continues the present form of sludge disposal.

Subsection 9.3 presents a payout analysis of four (4) alternatives considered favorable for consideration by the City of St. Petersburg.

9.2   USER FEE ANALYSIS

In this section, user fees are calculated for each of the most favorable alternatives. User fees or tipping fees are the price per unit of sludge for disposal of sludge. In Tables 9-1 through 9-6 ranges in user fees have been generated to cover a low estimate and a high estimate. Generally, the range is based on the uncertainty in projecting revenues and costs. For some alternatives, it is based upon a range of grant funding. In any case, the user fee is an equitable way to compare the cost for each favorable alternative.

A. Alternative B - Multiple-effect Drying/Full Service Operator/ Distribution and Marketing

1. Description

The Carver-Greenfield process is a drying technique utilizing the basic principle of evaporation. Successive stages of evaporation are accomplished by using the hot steam produced in the later stages of evaporation under decreasing pressure. An oil is added to the sludge in the process to keep the sludge "fluid" as it dries in the later stages. This oil is captured in the last stage plus any sewage oil that may be in the sludge, and recycled. Facilities in this treatment process are more complicated and capital cost intensive than certain other alternatives, such as lime encapsulation. However, it is a proven technology when operated by trained personnel. In a full service application, this alternative is acceptable.

Additional features of the process include: the ability to handle scum and septage as well as the sludge, lower energy usage through steam reuse, and possible sale of steam for energy recovery. Revenues from the sale of the end product, a 95-98% dry solids soil conditioner, would be shared with the City of St. Petersburg in this full service operation.

2. Projected Disposal Costs

This cost is referred to as user fees and is derived when revenue sales, and other income revenue, are subtracted from the total operating costs and financing costs. The user fee was determined to be a range of $110 to $150 in 1987 and a range of $180 to $340 in 2007 as presented in Table 9-1.

B. Alternative E - In-vessel Composting/Full Service Operator/ Distribution and Marketing

1. Description

While many of the full service composting systems differ in detail, the basic process and costs of operation do not tend to vary greatly. Accordingly, two in-vessel systems with proven technology and previous full service experience are discussed below, and the costs can be considered representative of this approach regardless of the commercial system used. The two systems described below are: (1) American Bio-Technology; and (2) Taulman-Weiss.

American Bio-Technology

The system, marketed under the Air-Lance trade name, consists of an enclosed reactor into which are placed slotted vertical stainless steel tubes, much like long well points, through which air is pumped into the sludge/bulking agent mixture. Exhaust blowers

TABLE 9-1

ALTERNATIVE B

PROJECTED DISPOSAL COSTS

MULTIPLE-EFFECT DRYING/FULL SERVICE OPERATOR

($ x 1,000)

Description	1987		1997		2007	
	Low	High	Low	High	Low	High
Sludge (tons DWS)	5,372	5,372	5,934	5,934	6,556	6,556
User Fee ($/ton DWS)	107.98	209.42	141.16	259.52	184.56	340.15
Revenues						
User Fees	580	820	840	1540	1,210	2,230
Sales	390	270	780	540	1560	1080
Other Income	250	125	200	75	50	50
Total Revenues	1,220	1,520	1,820	2,160	2,820	3,360
Total Operating Costs	700	860	1,300	1,500	2,300	2,700
Net Revenues	520	660	520	660	520	660
Financing Costs	520	660	520	660	520	660
Rounded User Fee ($/ton DWS)	110	210	140	260	180	340

extract the air from the vessel thus creating a negative pressure and minimizing odor problems. The walls of the vessel are slanted outward toward the bottom thus minimizing hang-up problems.

The sludge in the front of the reactor is mixed with the bulking material and is fed by conveyor to the top of the primary reactor. Composted sludge is extracted out of the bottom of the reactor and conveyed to an enclosed curing reactor, from where it is finally conveyed to finished storage. The entire plant is enclosed and will create minimal adverse environmental impact.

Taulman-Weiss

The Taulman-Weiss composting system is essentially the same as the American Bio-Technology process in that vertical silos are arranged in series, beginning with a reactor and ending with storage. In the Taulman-Weiss system air is pumped in at the bottom and exhausted at the top. A desirable feature of this system is the continuous monitoring of O_2 and CO_2 in order to provide the appropriate air flow. The recommended bulking agent and carbon source is sawdust, although other materials such as wood chips, leaves or shredded paper is adequate.

2.   Projected Disposal Costs

The projected disposal costs for a full service operator alternative include both the internal hauling costs and the product related operations costs. We have reviewed the previous full service operator quotations for the Orlando Sludge Management and Disposal Program and have solicited several quotations for this project from full service operators. These values were used as a check for the calculation presented on Table 9-2. The expected bids should be in the $120 to $210 per ton dry weight solids.

C.   Alternative G - City/County Incineration/Resource Recovery

1.   Description of Current Program

Alternative G includes the complete City/County incineration project. This project consists of contract hauling dewatered sludge cake from each of the City's regional sanitary facilities to a central sludge incineration facility located north of the Pinellas County Resource Recovery Facility at 28th Street. The sludge cake transportation would be contracted to a sludge hauler. The system would also include landfill gas recovery systems transporting the solid waste derived fuel to the incinerator. The sludge and grease would initially enter a receiving building in which the sludge would be conveyed to a sludge holding vessel. The oil and grease would be received, concentrated and stored and tranferred to vapor scrubbers prior to discharge into the feed system. Once the feed materials and fuel have been properly prepared, each would be transported to the fluidized bed incinerator. The incinerator facilities consists of two 24-foot

TABLE 9-2

ALTERNATIVE E

PROJECTED DISPOSAL COSTS

IN-VESSEL COMPOSTING/FULL SERVICE OPERATOR

($ x 1,000)

Description	1987		1997		2007	
	Low	High	Low	High	Low	High
Sludge (tons DWS)	5,372	5,372	5,934	5,934	6,556	6,556
User Fee ($/ton DWS)	121.00	208.49	139.87	246.04	193.72	308.11
Revenues						
User Fees	650	1,120	830	1,460	1,270	2,020
Compost Sales	270	190	540	370	1,070	730
Other Income	380	190	330	140	60	60
Total Revenues	1,300	1,500	1,700	1,970	2,400	2,810
Total Operating Costs	500	600	900	1,070	1,600	1,910
Net Revenues	800	900	800	900	800	900
Financing Costs	800	900	800	900	800	900
Rounded User Fee ($/ton DWS)	120	210	140	245	195	310

diameter unit until the year 1990 and an additional unit added to treat sludge production until the year 2000. Following incineration, the waste heat would be utilized to generate steam for transport to an on-site steam turbine and subsequent generation of electricity for sale to Florida Power Corporation.

2.   History of the Pinellas County Regional Program

The Pinellas County 201 Facilities Plan - Chapter 5, County-wide Sludge Disposal Study of 1978 called for a program of first dewatering the sludge and then drying it to generate a salable product. The County later prepared another study that proposed an incinerator to utilize Toytown landfill gas in burning the County's sludge and grease. This report, The Landfill Gas Utilization Study, was presented to the City of St. Petersburg's staff in September, 1983. The facilities were comprised of one 21-foot diameter incinerator for both the County's and City's sludge and the County-wide disposal of grease. In this study, the tipping fee was presented in the range of $140 to $150 per dry ton without grant funding. With the possible FDER grant funding initiated by the City of St. Petersburg, it was estimated that the tipping fee could be reduced to approximately $65 to $75 per dry ton. This economical and cost-effective program was strongly supported by the City.

In an August, 1984, report the County presented a incinerator program with a tipping fee of $105 per dry ton. This plan now recommends two 24-foot incinerator units for reliability. This draft facility plan amendment has not yet been accepted by the EPA.

Further, in December of 1984, a value engineering review of the proposed revised incinerator program was conducted. The VE report makes several recommendations for maximum savings among which are:

-     2 - 20-foot diameter incinerators in lieu of 2 - 24-foot units.

-     Installation of a 3.1 MW generator in lieu of 5.1 MW.

Although a total savings of $3,994,000 was reported, the total cost of the project is estimated in the VE Report at $20,900,000. This estimate includes many items missing from the original estimate, such as electrical, and instrumentation costs.

3.   Projected Disposal Costs

The projected disposal costs for this study included consideration of a projected "low" and "high" disposal cost as shown on Table 9-3. In the user fees for this alternative the "low" and a "high" estimate represent two different incinerator programs. The "low" estimate reflects the cost cited in the latest County report

TABLE 9-3

ALTERNATIVE G

PROJECTED DISPOSAL COSTS [a]

CITY/COUNTY INCINERATION PROJECT

($ x 1,000)

Description	1987 Low[b]	1987 High[c]	1997 Low	1997 High	2007 Low	2007 High
Sludge (tons DWS)	---	9,600	---	11,100	---	11,450
Grease (tons DWS)	---	2,900	---	3,000	---	3,050
User Fee ($/ton DWS)	---	313.60	---	313.47	---	325.51
Revenues						
User Fees	---	3,920	---	4,420	---	4,720
Resource Recovery	---	1,300	---	1,500	---	1,700
Total Revenues	---	5,220	---	5,920	---	6,420
Total Operating Costs[d]	---	2,500	---	3,200	---	3,700
Net Revenues	---	2,720	---	2,720	---	2,720
Debt Service	---	2,470	----	2,470	---	2,470
Coverage (Minimum)	---	1.10	---	1.10	---	1.10
Coverage (Actual)	---	1.10	---	1.10	---	1.10
Reserves	---	250	---	250	---	250
Rounded User Fee[d] ($/ton DWS)	130	315	155	315	185	325

NOTES:
(a)   Both low and high values based on $6,380,000 grant contribution.

(b)   Based on County values presented, a 17% DWS sludge, and two - 24-foot diameter incinerators to the year 1990 and one more to the year 2000. Financing is by bond issue.

(c)   Based on estimates from County reports. Debt service is based on estimated project cost of $20,900,000 as determined for maximum savings in the Value Engineering Study.

(d)   Includes cost for internal hauling.

to EPA. It is based on their cost estimate of two 24-foot diameter incinerator units until the year 1990 demands. The total construction cost estimate is $19,852,000 of which $6,380,000 will be grant funded. The financing arrangement assumed is by municipal bond issue, which is reasonable for the degree of capital out-lay which this program represents.

Using the figures of $105/ton @ 12,500 tons sludge and grease, total User fee revenues are $1,312,000. Revenue from other sources are $1,304,000, for a total revenue of $2,616,000. Yearly O & M is projected to be $3,099,000 for a net loss of $483,000. Insufficient information is available to trace or correct the stated figures.

The "high" estimate is based on the VE program for maximum savings and current engineering practice estimates for technical and engineering cost. The total construction cost estimate is $16,254,000. The total project estimate is $20,900,000, of which $6,380,000 (approximately 30%) will be grant funded. The financial analyses in Table 9-3 is based on a municipal bonding arrangement with a maximum 1.15 coverage.

As a result of the above analysis, ranges in the project disposal costs were determined for FY 1987, FY 1997 and FY 2007. Assuming that both user fees for the City and the County sludges would be equal and that the user fees for the County grease would also be the same, unit values were determined in dollars per ton dry weight solids (DWS). The above assumption in no way precludes the future determination of appropriate user fees for the respective entities.

In summary, the proposed disposal costs were estimated as follows: 1987-$105 to $320 per ton DWS; 1997-$130 to $320 per ton DWS; and 2007-$160 to $330 per ton DWS. Note that the increasing value of products, such as electricity, have offset the projected increases in operational costs for the "high" estimate.

D.   Alternative H - Lime Encapsulation/City Only/Hauling FOB

　　1.   Description

Lime has been used for stabilizing sludges for many years. The additions of $Ca(OH)_2$, hydrated lime, in sufficient quantities results in the elevation of the pH and subsequent disinfection and odor control. Unfortunately, this represents a short-term solution since in time the pH of the sludge-lime mixture drops and the odor problems resume. Additionally, the mixing of sludge and $Ca(OH)_2$ does not result in a reduction of the water content.

These problems with lime stabilization can be solved by using quicklime (dehydrated lime or CaO) instead of hydrated lime. The quicklime reacts with the water in the sludge as:

$$CaO + H_2O \qquad Ca(OH)_2 + Heat$$

The heat produced assists in the disinfection of the sludge and dewatering occurs because of the uptake of water in the reaction. Sludge which has been treated with a sufficiently high dosage of quicklime will be dry and almost sterile. In addition, some of the water is evaporated due to the elevated temperatures.

Since the quicklime reacts with the water, a sludge with a low solids concentration (high moisture) will require considerable amounts of lime.   On the other hand, treating an already mechanically dewatered sludge with quicklime presents problems with mixing and distribution of lime in the sludge.

These problems have been successfully addressed in a system marketed by Roediger Pittsburg, Inc. named Lime Post Treatment (LPT). This system uses a patented paddle mixer for achieving adequate contact between sludge solids and the quicklime by first grinding the sludge into small pellets and then powdering them. These pellets are thus dehydrated on the surface and do not stick together, even after long-term storage. The lime also penetrates into the core of the sludge particles, thus achieving disinfection. Both the pH rise as well as the elevated temperature is responsible for the destruction of pathogens.

The actual system is fairly simple, consisting only of a storage silo for the quicklime, the paddle mixer and associated conveyor.

2.   Estimated Disposal Costs

The estimated disposal costs are derived as follows:

Assumption 1: Product is sold for $20 per ton, escalated at 7%. Since 1 ton CaO produces 1.32 tons, Ca(OH)$_2$, total tons of finished product on a dry basis is dry solids + 1.32 CaO.   The application rate of CaO is 0.25 ton per dry ton of solids.

	1987	1997	2007
Sludge (dryton/yr)	5,372	5,934	6,556
Dry Product (ton/yr)	7,140	7,890	8,720
Revenue, $20/dry ton escalated at 7% annually	$ 164,000	$355,000	$776,000

Assumption 2:   Product is given away at treatment plant. No cost.

Assumption 3:   Product is disposed of by truck. Haul costs for encapsulated sludge as the costs of hauling 30% dewatered sludge (Alternative J). Based on a specific gravity of 1.14, cubic yards of   product   equals   3.47   times   the   dry   tonage.

	1987	1997	2007
Cubic Yards of 30% DWS Product	24,800	27,400	30,300
Contract Haul price,$/yd^3	11.50	24.80	53.50
Total Contract Haul Cost $/year	$ 285,200	$ 679,500	$ 1,621,000

Due to the range of financial possibilities with this alternative and the untried nature of the possible market in the City of St. Petersburg, we recommend that a pilot scale testing program be instituted to ascertain the possible success of Alternative H. Table 9-4 presents the wide range in possible disposal costs and illustrates that Alternative H, if successful, would be the least cost program for the City of St. Petersburg at $25 to $110 per dry ton in 1987, $40 to $215 per dry ton in 1997 and $80 to $440 in 2007.

If the lime encapsulation pilot program is implemented, the following are some suggestions for further investigations:

-       Reduce the sludge digestion time with stabilization being mainly accomplished by the liming process.

-       Seek EPA Innovative or Alternative status.

-       Utilize the product for City of St. Petersburg fertilizer needs.

-       Analyze markets for lime encapsulated sludge.

-       Make Application for FDER construction grant funding.

E.   Alternative I - In-vessel Composting/City Ownership and Contracted Full Service Operations/Distribution and Marketing

    1. Description

        Alternative I provides for City ownership and contracted full service operations and marketing for an In-Vessel Composting system.   This alternative includes the opportunity for incorporation of possible State of Florida and/or U.S.E.P.A. construction grants funding for the project at an assumed 55% level. A higher funding level may be obtained if it qualifies for Innovative or Alternative funding. The City would procure and local share fund a "turn-key" type design/build composting facility located at the Southwest Wastewater Treatment Plant on a competitive basis. The facilities provided would be the same as those for Alternative E.

TABLE 9-4

ALTERNATIVE H

PROJECTED DISPOSAL COSTS

LIME ENCAPSULATION

$ (x 1,000)

Description	1987		1997[a]		2007[a][b]	
	Low	High	Low	High	Low	High
Sludge (tons DWS)	5372	5372	5934	5934	6556	6556
User Fee $/dry ton DWS	23.83	107.41	40.95	215.20	77.33	442.80
Capital Cost[c] (including contingencies and fees)	96	96	171	171	337	337
O & M Costs						
Quicklime	117	117	280	280	669	669
Labor 75	75	75	140	140	263	263
Power	4	4	7	7	14	14
(Sale) Disposal	(164)	285	(355)	679	(776)	1,620
Net Annual Cost	128	577	243	1,277	507	2,903
User Fee Rounded (tons DWS)	25.00	110.00	40.00	215.00	80.00	440.00

NOTES:

(a) Capital costs FY 1997 and 2007 based on escalated cost to replace equipment and 3% engineering cost.

(b) Salvage value of equipment is neglected.

(c) Annual cost of the total capital cost from Table 7-12 ammortized over 12 years @ 10%.

The responsibility for permitting, operation and marketing/sales/disposal of the compost product would be borne by the full service operator. A complete description of this alternative is provided in Section 7. The alternative analyses and discussion of Alternative I cost and non-cost factors are presented in Section 8.

2.   Estimated Disposal Costs

The estimated disposal costs for Alternative I are presented on Table 9-5. The user fee or cost per ton dry weight solids is presented as an aggregate of both the necessary cost of operations which would be competitively procured and the cost of the local share of the capital requirements. The "low" estimate reflects 65% Federal/State grant participation and good sales and operations conditions. The "high" estimate reflects 55% Federal/State grant participation and poorer sales and operations based upon various factors as described in Section 8.

F.   Alternative J - Contract Hauling/Land Application

1.   Description

This alternative combines two common components of sludge management: (1) contracting; and (2) hauling of sludge to a land application site. In this alternative the company contracted to do the hauling would conduct all phases of disposing of the sludge. The contracted hauler would provide the trucks to collect the dewatered sludge at each of the wastewater treatment plants as well as secure the required permits, if any, to dispose of the sludge. All the equipment needed for land application would also be provided and the personnel needed in the total operation would be retained by the contracted firm. The contracted hauler would be paid by the city based on a unit of sludge (i.e., per cubic yard) and still, in turn, charge the grove, farm, or other type land owner receiving the sludge. Some haulers sell the sludge on the basis of nitrogen and phosphorus content.

A consideration for managing contract hauling is the multi-year contract approach as discussed in Subsection 6.4,C. Using a long-term contract for hauling with an annual escalation allowance would save the City the cost of yearly procurement. Also, it would allow them to make better projections for annual costs for the Public Utilities Department.

Section 7 provides a more detailed description of this alternative and describes the effect of chemical quality of the sludge on possible marketability. Such non-cost factors as environmental effects and aesthetics are addressed in Section 8. This section will compare this alternative mainly on projected costs.

2.   Disposal Costs

TABLE 9-5

ALTERNATIVE I

PROJECTED DISPOSAL COSTS

IN-VESSEL COMPOSTING WITH CITY OWNERSHIP

($ x 1,000)

Description	1987		1997		2007	
	Low	High	Low	High	Low	High
Sludge (tons DWS)	5,372	5,372	5,934	5,934	6,556	6,556
User Fee ($/ton DWS)	161.95	228.96	163.46	256.15	173.89	305.06
Revenues						
User Fees	870	1,230	970	1,520	1,140	2,000
Compost Sales	270	190	540	370	1,070	730
Other Income	30	0	30	0	30	0
Total Revenues	1,170	1,420	1,540	1,890	2,240	2,730
Total Operating Costs	500	600	900	1,070	1,600	1,910
Net Revenues	670	820	640	820	640	820
Debt Service	510	650	510	650	510	650
Coverage (Min.)	1.25	1.25	1.25	1.25	1.25	1.25
Coverage (Actual)	1.31	1.26	1.25	1.26	1.25	1.26
Reserves	160	170	130	170	130	170
Rounded User Fee ($/ton DWS)	160	230	165	255	175	305

In this alternative, the disposal cost is based on the fee paid to the hauler per cubic yard of dewatered sludge. Table 9-6 lists the cost for disposal for various percentages of dewatered sludge for the years shown, since the final sludge solids content is as yet undetermined. The charge is escalated by 8 percent in this alternative due to the assumed escalation of fuel to be also 8 percent. Fuel is important in determining the charge by a hauler and therefore will more closely follow its increase. The 1985 charge of $9.90 reflects the increases due to the new FDER Regulation 17-7, Part IV.

In this cost analysis, it is seen that the contract hauling by trucking will be between $48 and $74 for the year 1985 and will increase at the projected compounded rate of inflation. Not only does this alternative meet many non-cost criteria but it is also an economical sludge disposal method.

G.   Summary

The preceding six (6) Subsections 9.2, A through F, present the most favorable alternatives for consideration. Table 9-7 summarizes these alternatives. It may be noted that generally the full service operator alternatives range in estimated sludge disposal costs of $120 to $240 dollars per ton dry weight solids. The City/County regional incinerator project has a projected disposal cost of from $105 to $320 dollars per ton dry weight solids. The contract hauling alternative is escalated to fiscal year 1987 and has projected cost of from $55 to $85 per ton dry weight solids. The lime encapsulation alternative has a disposal cost if the material cannot be marketed of from $70 to $110 per ton dry weight solids in 1987. If this product can be marketed in the City of St. Petersburg, at $25/ton, this is the most cost-effective alternative.

9.3   PAY-OUT PERIOD ANALYSES

A.   Description

The payout period is most commonly defined as the length of time to recover the first cost of an investment from the net cash flow produced by that investment for a zero percent interest rate. In this report, the method is used to compare the cost effectiveness of various alternatives compared to the least cost alternative of contract hauling. In this analyses the number of years it takes to payback the deficit created in the first years, by selecting an alternative initially more expensive than contract hauling, is the "payout period".

We believe that the maximum payout period acceptable to existing alternatives is 10 years. Certain Florida municipalities have financial guidelines recommending capital investments for projects having payout periods from 5 to 10 years. Longer periods may not be financially favorable in terms of budget and rate planning for the utility.

B.   Results

TABLE 9-6

ALTERNATIVE J - CONTRACT HAULING/TRUCK

PROJECTED DISPOSAL COSTS

Year	Sludge Production (Dry Tons/Day)(a)	Dewatered Sludge Production (Cubic Yards/Day)(a)			$ Cost/ Cubic Yard(b)	Total Cost Per Dry Ton(d)		
		15% DWS(c)	18% DWS	21% DWS		15% DWS	18% DWS	21% DWS
1985	14.6	109	90	71	$ 9.90	$ 74.00	$ 61.00	$ 48.00
1987	14.9	110	93	76	11.50	85.00	72.00	59.00
1997	16.3	122	102	82	24.80	186.00	155.00	125.00
2007	18.4	137	114	91	53.50	398.00	331.00	265.00

NOTES:

(a) Per calendar day.
(b) Cost escalated by 8% yearly.
(c) DWS-Dry Weight Solids
(d) Calculated by: (Cubic Yards of Dewatered Sludge Per Day x Cost Per Cubic Yard) ÷ Dry Tons Per Day = Total Cost Per Dry Ton (rounded off to the nearest dollar).

TABLE 9-7

SUMMARY OF FAVORABLE ALTERNATIVES(a)

Alternative	Description	Estimated Project Cost	Estimated O & M Cost	Estimated Disposal Costs 1987 Low	1987 Avg.	1987 High	1997 Low	1997 Avg.	1997 High	2007 Low	2007 Avg.	2007 High
B	Multiple-effect Drying Full Service Operator	-	-	$110	$160	$210	$140	$200	$260	$180	$260	$340
E	In-Vessel Composting Full Service Operator	-	-	120	165	210	140	190	245	195	250	310
G	Regional Sludge Incinerator City/County Project	(b)	(b)	130	220	315	155	235	315	185	255	325
H	Lime Encapsulation(c) City Owner/Operation	$0.44M $1.17M(d)	$ 0.18M	25	70	110	40	125	215	80	260	440
I	In-Vessel Composting City/Contract Operations(e)	$ 8.5M	$ 0.55M	160	195	230	165	210	255	175	240	305
J	Hauling & Disposal City Contracting	-	-	60	70	85	125	155	185	265	330	400

(a) All estimates based on assumptions and costs discussed in Sections 7 and 9 of this report. No costs for dewatering are included.

(b) Costs listed are FY 1985. HDR projections (Reference No. 23) are $19.9 M and $3.10 M, and values from DRMP analyses is $20.9 M and $2.5 M for project and O & M cost, respectively.

(c) Costs listed are FY 1985.

(d) The $1.17 M is the escalated 1997 capital cost of a new process equipment plus 3% for engineering when the service life of 12 years for the original equipment has expired. No grant funding is included.

(e) Costs listed are FY 1987.

Figures 9-1 through 9-4 show the results of the payout analysis. Figure 9-1 demonstrates how the annual user costs shown on Table 9-8 changes with time. The main information to be drawn form Figure 9-1 is Alternative J - Contract Hauling and Alternative Hare the least annual cost alternatives. However, Alternative J becomes a lesser cost effective alternative in later years. This is primarily due to higher fuel costs and longer hauls.

In Figure 9-2 is shown the payout period for Alternative G-1. Both of Alternatives G-2 and G-3 would not payout in the 20 year study period. The best possible senario, Alternative G-1, pays back in 15 years. This is longer than what is recommended. If the base user fee is reduced to $80 per dry ton (not including internal hauling), this alternative would payout in the 10 year time limit.

The payout analysis for Alternative E - in-vessel composting is shown on Figure 9-3. The alternative is not considered cost effective due to the 20 year payout period. If the alternative user fee could be reduced to $105 per dry ton (including internal haulings) it would payout inthe 10 year limit. Alternative B -Multiple-effect drying has the same limitations.

The Alternative H - Lime encapsulation payout analysis is shown on Figure 9-4. The user fee represented is the average value. No revenue is considered for the produt and neither is a hauling charge. Under these circumstances, the alternative would payout in only three years.

9.4   RECOMMENDATIONS

Based on the user fee and pay-out period analyses, we recommend that the City pursue the City/County Incineration project as long as federal funds are awarded and that favorable terms can be negotiated with Pinellas County. In the interim until the incineration project can be placed in operation we recommend that the City continue its sludge hauling contracts. Concurrently, we recommend that a pilot scale Lime Encapsulation project be implemented by the City of St. Petersburg such that the project costs and potential revenues can be more accurately established.

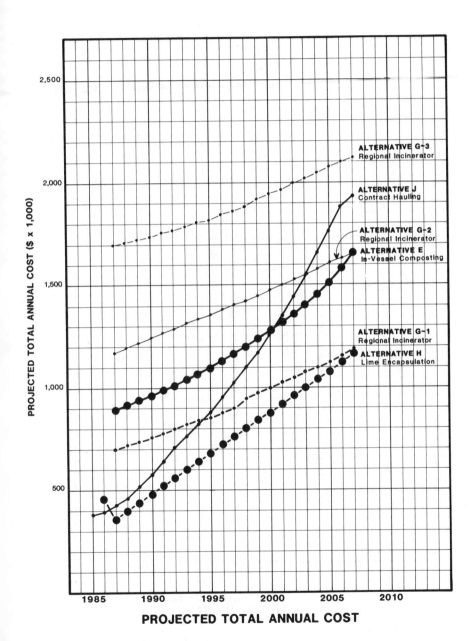

PROJECTED TOTAL ANNUAL COST

FOR ALTERNATIVES E,G,H & J

Figure 9-1

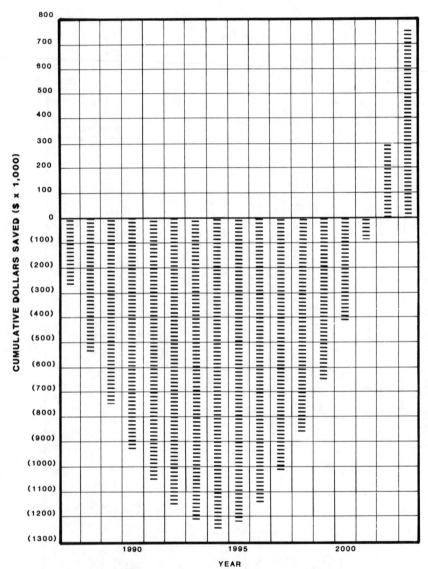

**PAYOUT PERIOD FOR REGIONAL INCINERATOR
VERSUS CONTRACT HAULING**

FOR ALTERNATIVE G-1

Figure 9-2

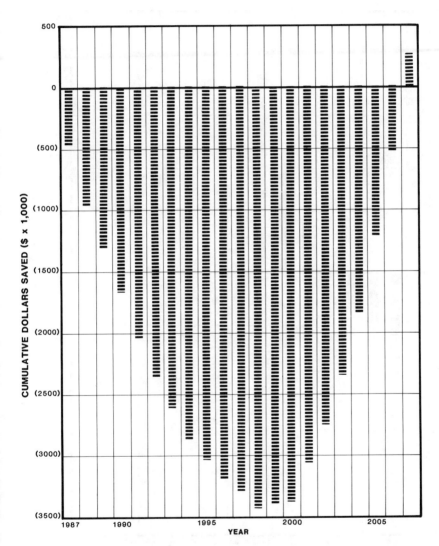

**PAYOUT PERIOD FOR IN-VESSEL COMPOSTING FSO
VERSUS CONTRACT HAULING**

FOR ALTERNATIVE E

Figure 9-3

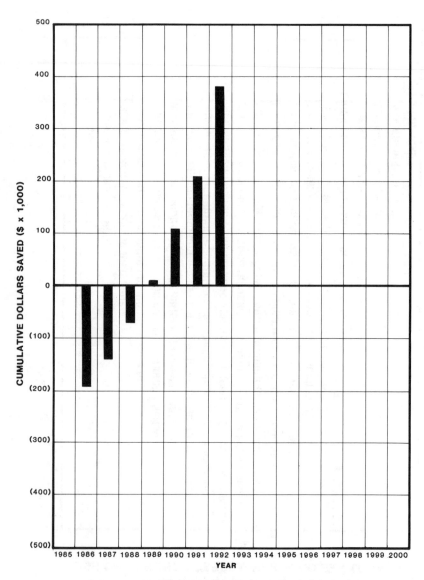

**PAYOUT PERIOD FOR LIME ENCAPSULATION
VERSUS CONTRACT HAULING**

ALTERNATIVE H

Figure 9-4

TABLE 9-8

ANNUAL COSTS USED FOR PAYOUT
PERIOD ANALYSIS[a]
(x $1,000)

Year	Alternative E In-Vessel Composting/ FSO[b]	Alternative G Regional Incinerator (Pinellas County)[b]			Alternative J Contract Hauling[c]	Alternative H Lime Encapsulation (Experimental)
		1	2	3[f]		
1985	---	---	---	---	390[e]	---
1986	---	---	---	---	400	590[g]
1987	890	700	1180	1700	430	360
1988	910	720	1200	1710	460	400
1989	935	740	1220	1730	520	440
1990	950	760	1250	1740	580	480
1991	980	780	1270	1760	640	520
1992	1010	800	1290	1770	705	560
1993	1040	820	1320	1790	760	600
1994	1070	840	1340	1810	820	640
1995	1100	860	1360	1820	885	680
1996	1130	880	1390	1850	955	720
1997	1160	900	1410	1870	1025	760
1998	1195	950	1420	1890	1100	800
1999	1230	980	1450	1920	1175	840
2000	1270	1000	1480	1950	1260	880
2001	1315	1030	1500	1970	1350	920
2002	1360	1050	1530	2000	1445	960
2003	1410	1080	1550	2020	1550	1000
2004	1465	1100	1580	2050	1640	1040
2005	1525	1130	1610	2070	1760	1080
2006	1585	1150	1630	2100	1890	1120
2007	1650	1180	1660	2120	1930	1160

NOTES:

(a) Sludge Production x $/Ton DWS = $ Annual User Fee Cost. Does not include costs for dewatering. Costs rounded to nearest $5,000.

(b) The average annual cost includes costs for internal hauling.

(c) For 17% DWS sludge.

(d) Alternative H average cost excludes revenues, yet assumes the product will be removed from site at no cost.

(e) For 15% DWS Sludge for 1985 only.

(f) Alternative G-1 values reflect the latest County proposal (Reference No. 23). Alternative G-2 is an average of the G-3 and G-1 values. Alternative G-3 reflects the values derived from the Value Engineering report (Reference No. 22). All numbers take into account a $6,380,000 grant contribution and internal hauling costs.

(g) 1986 costs do not reflect revenue to allow establishment of market.

SECTION 10

CONCLUSIONS AND RECOMMENDATIONS

10.1   SUMMARY OF THE REPORT

This report is the result of a thorough investigation of all available sludge disposal options for the City of St. Petersburg's Ultimate Sludge Disposal Plan.   Every sludge treatment technology was reviewed and selected processes were analyzed in a preliminary screening.   Various management approaches to sludge disposal were combined with the acceptable sludge treatment processes and disposal methods to form the sludge disposal alternatives for the City of St. Petersburg.   Section 8 utilizes the scientific method of transformation curve analysis to determine the most favorable alternatives.   In Section 9, each of these most favorable alternatives are then reviewed for short and long term user costs to determine which alternative should be selected.   The resulting list of conclusions and implementation schedule following this comparative analysis is in Section 10.

10.2   CONCLUSIONS

A.   Background

The City of St. Petersburg 201 Facilities Plan, written in 1978, recommended an alternative sludge handling and disposal methods in anticipation of:
".   The Toytown Landfill will discontinue operation in 1981.

.   The existing permit for the sod farm will expire June 1, 1980.

.   The sod farm experiences high water tables during the wet season because of an ineffective underdrain system.

.   Any drainage (surface or subsurface water drawn off to maintain a low water table) would have to be treated (probably biological and chlorination) before it could be discharged to a receiving water.

.   Even at a high sludge application rate (e.g. 30 dry tons per acre per year), 215 acres of land would be required to accommodate the year 2000 sludge production.   Consequently, the 50 acres presently available could only handle 23% of the future sludge production.

.   Potential virus contamination of ground water would require sludge sterilization prior to application.

.   Future regulations would probably limit agricultural related applications (e.g. forage crop feeding to beef or dairy cattle)."

Presented in the report was a list of disposal alternatives that were recommended for further consideration (in approximate order of priority):

".   Disposal at the Pinellas County Resource Recovery Facility.

.   Disposal at the Pinellas County Landfill.

.   Disposal at the Largo sludge pelletizing facility."

Sludge dewatering using belt filter presses has been recently implemented at the City's four wastewater treatment plants, but problems have arisen with the 201 Facilities Plan recommended disposal options.

The Pinellas County Resource Recovery Facility was not designed to handle the 15 to 21% weight solid sludge cake which is produced by the City's belt filter press operation. Also, the Pinellas Countywide Sludge Study recommended the closing the Toytown Landfill due to groundwater instead of another sludge-only landfill in the vicinity of the Resource Recovery Facility. In a Pinellas County-wide 201 Facilities report, an endorsement was given to a central drying facility that would produce a fertilizer-type sludge material. The Largo sludge pelletizing facility would be used on an interim basis. It was further recommended in this report that Federal grant assistance be obtained for the project.

In a subsequent report prepared by the firm of Henningson, Durham & Richardson, Inc., a plan was derived to burn sludge and grease in a County owned and operated fluidized bed incinerator. The facility would use landfill gas as a supplementary fuel supply and recover the waste heat to predry the sludge and generate electricity. This plan is a regional sludge disposal system in which the County has proposed to assume all capital requirements and, in turn, charge the City of St. Petersburg a sufficient tipping fee to both recover costs of operation and retire capital costs.

In order to lower the calculated tipping fee of this sludge disposal plan, a request for construction grant inclusion was made for the fiscal year 1985. Both the City and the County applied for the grant. The approval for inclusion in the Construction Grant Program was given to the County by the State of Florida Bureau of Wastewater Management and Grants.

B.   Conclusion

1.   With a projected 18 dry tons per day of wastewater in the year 2007 and the closure of nearby landfill sites and lack of nearby land application sites, the City of St. Petersburg requires an alternative long-term plan for sludge disposal.

2.   The sludge is of the quality that some form of land application or salable product would be appropriate.

3.   After complete review of all technologies, the ones selected as qualified for consideration are (Exhibits 4-1 through 4-13):

    -    Flash Drying
    -    Rotary Drying
    -    Multiple-effect Drying
    -    Solar Powered Drying

- Windrow Composting
- Aerated Pile Composting
- Earthworm Conversion
- In-vessel Composting
- Co-combustion
- Multiple Hearth Incineration
- Fluidized Bed Incineration
- Electric (Infrared) Furnace Incineration
- Lime Encapsulation

4. The long term sludge disposal alternatives must include all aspects of disposal from treatment to management approach to final disposal. Of all the alternatives reviewed, the ones found most favorable are:

    - Alternative B - Multiple-effect Drying/Full Service Operator/Distribution Marketing

    - Alternative E - In-vessel Composting/Full Service Operator/Distribution Marketing

    - Alternative G - Fluidized Bed Incinerator/City and County/Resource Recovery

    - Alternative H - Lime Encapsulation/City-only

    - Alternative I - In-vessel Composting/City Owner Operation

    - Alternative J - Contract Hauling/Land Application

5. After detailed costing analyses using a comparison of user fees and pay-out periods, the most favorable short term alternatives surveyed is:

Alternative	1987 User Fee
Contract Hauling	$70 Per Dry Ton
Lime Encapsulation	$70 Per Dry Ton

6. Participation of the City of St. Petersburg in the Pinellas County Regional Sludge/Grease Incinerator is dependent upon the County's agreement of a wholesale user cost for the City at $80/Dry Ton. This would make it competitive with contract hauling and lime encapsulation.

7. As hauling become less cost-effective and as soon as 1988, other favorable alternatives such as in-vessel composting and multiple-effect drying should be reviewed again.

8. This investigation had a limited scope excluding interlocal agreements with other entities which may compete with the favorable alternatives presented.

10.3   RECOMMENDATIONS

- Continue contract hauling as the primary method of sludge disposal through 1987. Revise hauling contract to a tri-annual basis.

- Experiment using the lime encapsulation process to better ascertain marketability and operational costs.

- Pursue the incineration alternative only if a tipping fee of less than $80/ton DWS is obtained on a long-term basis (over 10 years), and other terms are agreeable to the City.

- Investigate into hauling sludge to outside entities or municipalities for treatment, such as Manatee and Hillsborough County.

Depending on decisions made on the above recommendations:

- After agreement negotiations with Pinellas County and pilot testing of lime encapsulation, assess the favorability of other identified alternatives such as full/service operator procurement for either in-vessel composting or multiple-effect drying.

- Review the technology available in two (2) years and the current technology again to select a feasible sludge treatment process. This would be useful in such emerging U.S. application of technologies as in-vessel composting.

10.4 IMPLEMENTATION SCHEDULE

Presented in Table 10-1 is suggested implementation schedule with decision making milestones.

TABLE 10-1

IMPLEMENTATION SCHEDULE
CITY OF ST. PETERSBURG SLUDGE DISPOSAL PROGRAM

Activity	Date
Institute Tri-annual Hauling Contract	March, 1985
Conduct Full-scale Lime Encapsulation Pilot Test	March, 1985
(Apply for FDER Grant)	
Make Decision on Participation in Pinellas County Regional Sludge/Grease Incinerator	March, 1985
Make Decision on Lime Encapsulation	June, 1985
If Required, Re-evaluate Other Acceptable Alternatives Investigated in DRMP Ultimate Sludge Disposal Report	July, 1985
If Required a Request for Proposal for Full Service Operation	August, 1985
Implement Selected Alternative Sludge Treatment Study and Design	January, 1986
Start Construction of Selected Alternative	January, 1987
Renegotiate Sludge Hauling Contract	April, 1988
Start-up of Facilities for Selected Alternative	June, 1988

SECTION II

PROJECT UPDATE
MAY,1984 TO NOVEMBER, 1985

11.1 GENERAL

Sludge reuse has emerged in U.S. applications over the past two (2) years. The most significant advances have been in the application of composting technologies. As implementation problems are solved and installations are constructed, better operating and costing data has become available.

11.2 TECHNOLOGY UPDATE OF RECOMMENDED ALTERNATIVES

A.   Contract Haul and Dispose

Agricultural interests in Florida are better educated relative to accepting liquid and cake sanitary sludges. Grade I sludges per Florida Department of Environmental Regulation classification are readily accepted and are managed more appropriately. In recent West-Central Florida surveys, all concerns either accept the sludge, (applied in an acceptable manner as specified) either at no cost to the hauler or pay the hauler a slight fee. This situation has dampened expected increases in this alternative's costs.

B.   Lime Encapsulation

As a result of full-scale plant testing, several observations were made with regard to implementation.

1.   The process drives off an ammonia gas which is malodorous and attracts flies.

2.   A minimum of 15% by weight of quicklime dosage is required to attain a sustained pH of 12 or greater.

3.   Approximately 100% by weight of quicklime dosge is required to attain a sustained temperature of 160°F.

4.   ' Bacterial counts were reduced by lime disinfection at the 30% by weight quicklime dosage by three orders of magnitude.

5.   A total product percent solids of 30% could be attained with a 40 to 60% quicklime dosage. Natural drying is inhibited due to crust formation and clay-type of product consistency.

As a result of detailed testing appropriate potential applications include landfill material and intermittent rural agricultural needs.

C.   In-Vessel Composting

Numerous contracts of various types have been executed for in-vessel composting facilities nationally. Start-up and implementation problems are being solved. Compost markets are being established with a successful track record.

Appurtenant equipment is being manufactured to improve the cost-effectiveness of the technology. Refinements and value engineering inputs have significantly improved the competitive cost of composting for urban areas. We expect this trend to continue.

11.3 User Fee Update

Given a 20 ton per day design rate located in an urban area, as St. Petersburg, Florida, the following projected user fee ranges are presented for November, 1987:

- Haul and dispose contracting - $70 to 80 per dry ton of sludge.

- Lime Encapsulation - $100 to 110 per dry ton of sludge.

- In-Vessel Composting - $80 to 130 per dry ton of sludge.

We expect that both the haul and dispose contracting and the in-vessel composting alternatives to be competitive in life cycle costing. Finally, given state and/or federal grant assistance in-vessel composting may emerge as the most cost-effective and environmentally sound option for urban sludge utilization.

APPENDIX

FLORIDA DEPARTMENT OF ENVIRONMENTAL REGULATION

CHAPTER 17-7 REGULATIONS

CHAPTER 17-7

RESOURCE RECOVERY AND MANAGEMENT

PART IV: Domestic Sludge Classification, Utilization, and Disposal Criteria

17-7.50   Declaration and Intent.
17-7.51   Definitions.
17-7.52   Permit Requirements.
17-7.53   General Prohibitions for Domestic Sludges.
17-7.54   Criteria for Land Application or Disposal of Domestic Wastewater
           Treatment Sludge.
17-7.55   Criteria for Land Application or Disposal of Domestic Septage.
17-7.56   Criteria for land Application or Disposal of Food Service Sludges.
17-7.57   Criteria for Land Application or Disposal of Processed Domestic
           Sludges.
17-7.58   Criteria for Land Application or Disposal of Composted Domestic
           Sludges.
17-7.59   Effective Date.

PART IV: Domestic Sludge Classification, Utilization, and Disposal Criteria

17-7.50   Declaration and Intent.

The Florida Department of Environmental Regulation (department) is empowered to plan for and regulate the collection, transport, storage, separation, processing, recycling, and disposal of solid wastes including sludges, and to protect the air and waters of the state from pollution. The department is also charged with the promotion of recycling, reuse, or treatment of solid wastes.

The department finds and declares that the improper management and disposal of sludges contributes to air and water pollution and presents a threat to the public health, safety and welfare. The department also finds and declares that in order to regulate the management and disposal of sludges in a manner that is effective but yet not overly burdensome, regulations specific to management and disposal of sludge must be applied.

Therefore, it is in the intent of the department to regulate the management and disposal of domestic sludges generated by new and existing sources in a manner to assure protection of the environment and public health. The department also intends in this rule to encourage the proper recycling, reuse, treatment, or agricultural use of certain domestic sludges. Industrial sludges, industrial wastewater treatment sludges, air treatment sludges, and water supply treatment sludges are not regulated by this part.

The department intends to require resource recovery and management facility permits for sludge treatment, storage, and disposal facilities as directed by law. However, the department recognizes that special conditions exist for permitting sludge facilities, and by this rule intends to clarify when sludge sites qualify for exemptions from such permits as part of normal farming operations, and intends to establish a general permit rule for sludges that are not significantly contaminated.

Specific Authority: 403.061, 403.704, F.S. Law Implemented: 403.021, 403.061, 403.087, 403.701 through 403.715, F.S. History: New 6-16-84.

17-7.51   Definitions.

The following words, phrases or terms as used in this part, unless the context indicates otherwise, shall have the following meaning:

  (1)   "Aerosol" means airborne droplets or particles.
  (2)   "Agricultural Lands" means all lands being used for agricultural purposes.
  (3)   "Department" means the Florida Department of Environmental Regulation.
  (4)   "Composted Domestic Sludge" means sludge that has been composted at or above 55°C for three consecutive days in a mechanical composter or an aerated pile, or at or above 55°C for fifteen consecutive days in a windrow and at least five turnings of the windrow.

(5)   "Conservation Plan" means a formal document, prepared or approved by a local Soil and Water Conservation District Board organized pursuant to Chapter 582, Florida Statutes, which outlines a system of management practices to control soil erosion, reduce sediment loss or protect the water quality on a specific parcel of property.

(6)   "Dewatered" means a content of 12 percent or greater solids (by weight) in a sludge.

(7)   "Disinfection" means the selective destruction of pathogens in wastewater effluent or sludge as described in Chapter 7 of EPA 625/1-79-011, "Process Design Manual for Sludge Treatment and Disposal." This manual is adopted and made a part of this rule by reference. A copy of this document may be obtained by writing the department, and may be inspected at all DER offices.

(8)   "Domestic Septage" means all solid wastes containing human feces, or residuals of such, which have not been stabilized or disinfected. Not included are food service sludges and industrial sludges.

(9)   "Domestic Sludge" means a solid waste resulting from sewage, septage, or food service operations, or any other such waste having similar characteristics. Domestic sludge may be liquid, semisolid, or solid but does not include the treated effluent from a domestic wastewater treatment plant.

(10)   "Domestic Sludge Disposal" means the disposal of domestic sludge in accordance with disposal criteria presented in Part IV, Chapter 17-7, Florida Administrative Code.

(11)   "Domestic (Sewage) Wastewater Treatment Sludge" means any sludge generated by a domestic wastewater treatment plant.

(12)   "Dried Sludge" means sludge that contains more than 89 percent solids by weight.

(13)   "Dry Weight" means the weight measured after oven drying at $103^\circ - 105^\circ C$ to constant weight.

(14)   "Food Service Sludges" means oils, greases, and grease trap pumpings generated in the food service industry.

(15)   "Human Food Chain Crops" means all crops which may be harvested for human consumption; all crops which may be fed to animals which may be consumed by humans.

(16)   "Incorporation into the soil" means the injection of sludge beneath the surface of the soil or the mixing of sludge with the surface soil.

(17)   "Industrial Sludges" means all sludges that are primarily composed of materials generated through a manufacturing or other industrial process.

(18)   "Land Reclamation" means the restoration of productivity to lands made barren through processes such as erosion, mining or land clearing.

(19)   "Licensed Septage Hauling Service" means a person or service licensed to haul septage by the Florida Department of Health and Rehabilitative services.

(20)   "Lime Stabilization" means the addition of sufficient

quantities of lime to raise and maintain a sludge at a pH of 12.0 for two hours with such process being as described in Chapter 6 EPA 625/1-79-001, "Process Design Manual for Sludge Treatment and Disposal." This document is adopted and made a part of this rule by reference. A copy of this document may be obtained by writing the department, and may be inspected at all DER offices.

(21)   "Liquid Sludge" means any sludge that flows freely through piping, pumps, and liquid conveyance, transport, and spreading systems.

(22)   "Pathogens" means disease producing organisms.

(23)   "pH of Sludge-Soil Mixture" means the value obtained by sampling the soil to the depth of sludge placement and analyzing by the electrometric method. ("Methods of Soil Analysis, Agronomy Monograph No. 9, "C.A. Black, ed., American Society of Agronomy, Madison, Wisconsin, pp. 914-926, 1965.") This manual is adopted and made a part of this rule by reference. A copy of this document may be obtained by writing the department, and may be inspected at all DER offices.

(24)   "Processed Domestic Sludge" means any domestic sludge that has been stabilized, disinfected, dried, and also pelletized or granulated, and distributed as a soil conditioner or fertilizer constituent.

(25)   "Restricted Public Access" means that an area is not easily accessible to the general public.

(26)   "Root Crops" means plants with edible parts which parts are grown below the surface of the soil.

(27)   "Sludge " means a solid waste pollution control residual which is generated by any industrial or domestic wastewater treatment plant, water supply treatment plant, air pollution control facility, septic tank, grease trap, portable toilet or related operation, or any other such waste having similar characteristics. Sludge may be a solid, liquid, or semisolid waste but does not include the treated effluent from a wastewater treatment plant.

(28)   "Sludge Disposal Facility" means all contiguous land and structures, other appurtenances, and improvements on the land, used for sludge disposal or spreading of sludge on the land.

(29)   "Sludge Generator" means any facility that, as a normal function of its operation, produces a sludge. Residential septic tanks are excluded.

(30)   "Sludge Land Application Area Boundary" means the area of land used for sludge application.

(31)   "Sludge Land Application Area Boundary" means the outermost perimeter of the sludge land application area as it would exist at completion of the application activity.

(32)   "Sludge Pasteurization" means the heating of a sludge to 70°C for at least 30 minutes, or the heating of sludge to a temperature higher or lower than 70°C for a duration that produces a similar kill of microorganisms.

(33)   "Sludge Ponds and Lagoons" are containment areas containing only sludge.

(34)   "Stabilization" means the use of a treatment to render sludge or septage less odorous and putrescible, and to reduce the pathogenic content as described in Chapter 6 of EPA 625/1-79-011, "Process Design Manual for Sludge Treatment and Disposal." This manual is adopted and made a part of this rule by reference. A copy of this document may be obtained by writing the department, and may be inspected at all DER offices.

(35)   "Toxic Substances" means any of the following:
  (a)   Hazardous wastes as defined in Chapter 17-30, Florida Administrative Code;
  (b)   Hazardous substances as defined in the Comprehensive Environmental Response, Compensation, and Liability Act of 1980, Pub. L, No. 96-510, 94 Stat. 2767;
  (c)   A pollutant as defined in Chapter 376, Florida Statutes;
  (d)   A substance which is or is suspected to be carcinogenic, mutagenic, teratogenic, or toxic to human beings, or are acutely toxic as defined in the Fla. Admin. Code 17-3.021(1), or
  (e)   A substance which poses a serious danger to the public health, safety, or welfare.

(36)   "Treatment" means the process of altering the character, or physical or chemical condition of the waste to prevent pollution of water, air or soil, to safeguard the public health or enable the waste to be recycled.

(37)   "Type I Facility" means a wastewater facility having a design average daily flow of 500,000 gallons per day or greater.

(38)   "Type II Facility" means a wastewater facility having a design average daily flow of 100,000 to, 499,999 gallons per day.

(39)   "Type III Facility" means a wastewater facility having a design average daily flow of 2,000 to, 99,000 gallons per day.

(40)   "Wastewater" means the combination of the liquid and water carried wastes from waste generation sites, together with any entrained surface and groundwaters which may be present.

(41)   "Water Saturated Soil" means soil in which the spaces between the grains are filled with water.
        Specific Authority: 403.061, 403.704, F.S. Law Implemented: 403.021, 403.031, 403.061, 403.087, 403.701 through 403.715, 403.72, F.S. History: New 6-16-84.

17-7.52   Permit Requirements.

(1)   No sludge disposal or land application site shall be constructed, operated, modified, maintained, or expanded without a currently valid permit from the department unless exempted pursuant to Fla. Admin. Code Rule 17-7.03(2).

(2) Land application of any of the following sludges pursuant to the requirements of Fla. Admin. Code Rule 17-7.54(4), is considered normal farming operations and thus exempt from permitting requirements pursuant to Section 403.707(2)(a), Florida Statutes;

    (a) Grade I domestic sludge as classified in Fla. Admin. Code Rule 17-7.54(4).

    (b) Grade I composted sludge as classified in Fla. Admin. Code Rule 17-7.54(4).

    (c) Domestic septage properly treated by lime stabilization, sludge pasteurization, or other process which produces similar kills of microorganisms.

    (d) Food service sludge properly treated by lime stabilization, sludge pasteurization, or other process which produces similar kills of microorganisms.

    (e) Grade I or II processed domestic sludge.

(3) Land application of Grade II domestic wastewater treatment sludge or Grade II composted domestic wastewater treatment sludge which must be applied pursuant to the requirements of Section 17-7.54(5) is not a normal farming operation and must be performed pursuant to a general permit as provided in Fla. Admin. Code Rule 17-4.64, unless that sludge is applied to land owned by the generator. Applicable general permit requirements are stated in Florida Administrative Code Rules 17-4.54 and 17-4.64.

(4) Any sludge land application area that does not fulfill the requirements of a general permit under Fla. Admin. Code Rule 17-7.54(5) shall require a solid waste disposal site permit from the department as required in Part I of Chapter 17-7, Florida Administrative Code.

Specific Authority: 403.061, 403.704, 403.814, F.S. Law Implemented: 403.061, 403.701 through 403.715, 403.814, F.S. History: New 6-16-84.

17-7.53   General Prohibitions for Domestic Sludges.

(1) Ocean disposal of sludges and disposal of sludges in any natural or artificial body of water including groundwater is prohibited by Fla. Admin. Code Rule 17-7.04(3)(b).

(2) Sludge ponds and sludge lagoons shall only be constructed for the temporary storage or temporary treatment of domestic sludge. Such ponds and lagoons shall only be constructed for the temporary storage or temporary treatment of domestic sludge. Such ponds and lagoons containing domestic sludge shall be designed and utilized so that a sludge residence time of less than 12 months is assured. Ponds or lagoons in use on the effective date of this rule for the permanent storage (final disposal) of domestic sludge shall be phased out of use or converted to temporary holding areas at time of permit renewal, but in no case later than five years subsequent to the effective date of this rule. Florida Administrative Code Rule 17-

4.245 shall apply to all domestic sludge ponds and lagoons containing domestic sludge.

(3)   All domestic sludges other than processed domestic sludge and composted domestic sludge are prohibited from use on playgrounds, parks, golf courses, lawns, hospital grounds, or other unrestricted areas where frequent human contact with the soil is likely to occur.

(4)   Sludges which are hazardous wastes, as determined pursuant to Chapter 17-30, Florida Administrative Code, shall only be managed according to the requirements of that chapter.

(5)   The prohibitions of Fla. Admin. Code Rule 17-7.04 apply except those prohibitions described in Fla. Admin. Code Rule 17-7.04(3)(d), (e), (f), and (g) or any prohibition in conflict with the rules established in this part.

(6)   No sludge may be disposed into a collection/transmission system without prior consent of the owner of that system.

(7)   No domestic sludge shall be disposed of or applied to the land except in accordance with the provisions of this rule. A permit pursuant to this chapter is required unless such sludge is used in normal farming operations or is disposed of on the generator's own property and is treated and handled in compliance with all applicable department rules.

(8)   Pursuant to Fla. Admin. Code Rule 17-2.620(2), no person shall cause, suffer, allow or permit the discharge of air pollutants which cause or contribute to an objectionable odor.

(9)   The spraying of liquid sludge shall be conducted so that the formation of aerosols is minimized by utilization of best management practices.
Specific   Authority: 403.061,   403.704,   F.S.   Law Implemented: 403.021, 403.061, 403.087, 403.702, 403.704, 403.705, 403.707, 403.708, F.S. History: New 6-16-84.

17-7.54   **Criteria for Land Application or Disposal of Domestic Wastewater Treatment Sludge.**

(1)   All domestic wastewater treatment sludge shall be disposed of pursuant to the requirements of Fla. Admin. Code Rule 17-7.54(6). However, such sludges that have been analyzed pursuant to Fla. Admin. Code Rule 17-7.54(2) and have been determined to be a Grade I or Grade II sludge pursuant to criteria specified in Fla. Admin. Code Rule 17-7.54(3) may be applied to the land without a permit or under a general permit according to the criteria provided in Fla. Admin. Code Rule 17-7.54(4) and (5).

(2)   Sludge meeting the following criteria shall be classified as Grade I or II.
(a)   The interval between reported sludge analyses used for sludge classification shall be no greater than three months for a Type I domestic wastewater treatment facility, six months for a Type II domestic wastewater treatment facility, and 12 months for a Type III domestic wastewater treatment facility.

(b)   Parameters to be analyzed:

Total Nitrogen   - % dry weight
Total Phosphorus   - % dry weight
Total Potassium   - % dry weight
   Cadmium - mg/kg dry weight
   Copper   - mg/kg dry weight
   Lead   - mg/kg dry weight
   Nickel   - mg/kg dry weight
   Zinc   - mg/kg dry weight
   pH   - standard units
Total Solids   - %

(c)   Analysis of additional parameters may be required by the department based on changes in the quality of the wastewater or sludge as a result of: new discharges to the treatment plant, changes in wastewater treatment processes or process efficiency, changes in the utilization and/or disposal of the sludge, the potential presence of toxic substances in the sludge, or other considerations. The department may determine that certain listed parameters no longer require analysis or may be tested for less frequently than prescribed above where monitoring data show insignificant changes in sludge quality.

(d)   Sampling points, minimum sludge sample numbers, and sampling intervals shall be specified in the wastewater facility operation permit. All samples for sludge classification shall be representative and taken after final sludge treatment but prior to utilization disposal.

(e)   Samples for sludge analysis shall be collected as follows: One "grab" sample shall be taken in a clean glass container or two separate days each week for two consecutive weeks. These samples (approximately one quart each) shall be split, composited, and preserved as taken, in three one-gallon glass containers fitted with Teflon-lined caps or stoppers and refrigerated at 4°C. One of the gallon containers shall be maintained at a pH of less than 2 with sulfuric acid, and one shall be maintained at a pH of less than 2 with nitric acid. The third one-gallon container shall not contain a preservative. Directly after the second week sampling period, the three composite samples shall be thoroughly agitated and a quart subsample removed from each for analysis and placed in a glass container with a Teflon-lined cap or stopper. The quart samples shall be clearly labeled, including the date of collection, and shall be packed with ice in a well-insulated shipping container and forwarded to an analytical laboratory so as to insure receipt within 30 hours of packing. However, if no organic analysis is being required by the department, the glass containers fitted with Teflon-lined caps or stoppers may be replaced by polyethylene containers.

(f)   Samples shall be analyzed in accordance with methods described in Section XXV Miscellaneous Analysis for Soils, Sediments and Sludges, Part B, Domestic Sludge Analysis Methodology of the department's "Standard Operating Procedures Manual". Part B, of Section XXV of this manual is adopted and made a part of this rule by reference. A copy of this document may be obtained by writing the Department, and may be inspected at all DER offices. Sampling and testing methods shall be done in accordance with the requirements of Fla. Admin. Code Rule 17-4.246.

(3)   After completion of required analyses pursuant to Fla. Admin. Code Rule 17-7.54(2), domestic wastewater treatment sludge shall be classified as Grade I, II or III. A sludge is classified as a Grade I sludge if concentrations of all of the parameters listed below are less than the limiting criteria for Grade I sludges and the sludge is stabilized, and in the case of composted domestic sludge, the sludge is disinfected. A sludge is classified as a Grade II sludge if concentrations of any of the parameters listed below fall within the criteria for Grade II sludges but no parameters have concentrations above Grade II criteria and the sludge is stabilized and in the case of composted domestic sludge the sludge is disinfected. A Grade I sludge will be reclassified as a Grade II sludge if the one-year moving average of analysis results exceeds the maximum levels for Grade I sludge or the results for any one analysis exceed the maximum grade I levels by 15 percent. A sludge is classified as Grade III sludge if concentrations of any of the parameters listed below are greater than any of the criteria established for Grade III sludges and the sludge is not a hazardous waste as defined by Chapter 17-30, Florida Administrative Code.

Chemical Criteria in mg/kg dry weight

Parameter	Grade I	Grade II	Grade III
Cadmium	≤ 30	30-100	> 100
Copper	≤ 900	900-3000	> 3000
Lead	≤1000	1000-1500	> 1500
Nickel	≤ 100	100-500	> 500
Zinc	≤1800	1800-10,000	>10,000

(4)   Criteria for Land Application of Grade I Domestic Wastewater Treatment Sludge.

Domestic wastewater treatment sludge analyzed pursuant to Fla. Admin. Code Rule 17-7.54(2) which meets the criteria specified in Fla. Admin. Code Rule 17-7.54(3) for a Grade I sludge, which has not been found to contain significant concentration of toxic substances, may be applied to the land if the following conditions are met.

(a) Application is limited to sod farms, pasturelands, forests, highway shoulders and medians, plant nursery use, land reclamation projects and soil used for growing human food chain crops (excluding root crops, leafy vegetables, tobacco, and vegetables to be eaten raw).

(b) Use on playgrounds, parks, golf courses, lawns, hospital grounds, or other unrestricted public access areas where frequent human contact is likely to occur is restricted to processed domestic sludge and composted domestic sludge.

(c) Grade I - Application Rates. Application rates (gallons or pounds per acre per year) of Grade I domestic wastewater treatment sludge are limited by the nitrogen content of the sludge. The maximum annual application rate of total nitrogen is 500 pounds per acre during any 12-month period. In no case shall more than six dry tons of sludge be applied per acre per year.

(d) Grade I - Site Requirements

1. Buffer Zone Requirements - The sludge land application area shall be located no closer than 3,000 feet from any Class I water body or 200 feet from any other natural or artificial body of water, except canals used for irrigation purposes or bodies of water completely within and not discharging from the site. The sludge land application area shall be located no closer than 300 feet from shallow private water supply wells as defined in Fla. Admin. Code Rule 17-7.02(46) and occupied buildings (residences, offices, manufacturing facilities, etc.). The sludge land application area shall be located no closer than 500 feet from shallow public water supply wells as defined in Fla. Admin. Code Rule 17-7.02(46). Processed domestic sludge and composted domestic sludge are exempted from the above specified buffer zone requirements.

2. Soil Requirements - The pH of the sludge-soil mixture shall be 6.5 or greater at the time of sludge application. When growing human food chain crops, testing of the pH parameter shall be done on a semi-annual basis, otherwise testing shall be done annually.

3. Water Table Requirements - A minimum unsaturated depth of two feet is required above water table level at the time of sludge application to the soil. Water table level shall be determined by observing the standing water level in a three foot depth hole dug within the preceding 24 hours on the area to be used, or by

measuring the water level in a monitoring well evacuated within 24 hours prior to measurement. No sludge land application may be conducted during rain storms or during periods in which surface soils of the sludge land application area are water saturated.

4.   Special Requirements

a.   Sludge land application area topographical grades must be eight percent or less.

b.   The sludge land application area and an area 200 feet wide adjacent to and exterior of the sludge land application area boundary shall contain no subsurface fractures, solution cavities, or sink holes.

c.   Restriction of public access - The general public shall be restricted from the sludge land application area for a period of 12 months after each application. Processed domestic sludge and composted domestic sludge land application areas are exempted.

d.   Florida water quality criteria and standards as provided in Chapters 17-3 and 17-4, Florida Administrative Code, shall not be violated as a result of land application of sludge. Water quality testing of Grade I sludge application areas may be required if the department determines that sludge application not conforming to this section is evident. If water quality violations are indicated, the site owner shall cease any further sludge land application and have water quality tests performed as the department requires.

e.   When applied to unvegetated soils, stabilized domestic sludge other than processed domestic sludge and composted domestic sludge shall be incorporated into the soil within 48 hours of application.

f.   Depth of Permeable Soil - A layer of permeable soil of more than 2.0 feet thickness shall cover the surface of the land application area.

g.   Pasture vegetation on which sludge has been spread shall not be fed to livestock for a period of 30 days from the last application of sludge.

h.   Sludge land application areas shall not be used for domestic animal grazing for a period of 30 days following the last application of sludge.

i. No human food chain crops except hay, silage, or orchard crops shall be harvested from a sludge land application area for a period of 60 days following the last application of sludge.

j. No sludge shall be applied directly onto those portions of human food chain crops which are to be harvested and removed from the sludge land application area except hay on properly cropped and raked pastureland.

k. Domestic sludges may not be used for the cultivation of root crops, leafy vegetables, tobacco, or vegetables to be eaten raw.

l. Permanent records of actual application areas and application rates must be kept. These records must be maintained by the site owner and the sludge land applicator for a period of five years, and must be available for inspection upon request by the department, The Department of Health and Rehabilitative Services, or Local Environmental Program.

Sludge Records shall include:

(i) Date of sludge shipment or application.

(ii) Weather conditions when delivered.

(iii) Location of sludge destination.

(iv) Amount of sludge applied or delivered.

(v) Grade of Sludge.

(vi) Sludge use agreements.

(vii) Area of land where sludge is going to be applied, location and designated future use of the land.

(viii) pH.

(ix) Water table.

(x) Topographic grades.

(xi) Vegetational status of sludge land application area.

(xii) Incorporation of sludge.

(xiii) Depth of permeable soil.

(e) Persons who haul or use less than three cubic yards of Grade I Sludge per month are exempt from the requirements of 17-7.54(4).

(5) Criteria for Land Application of Grade II Domestic Wastewater Treatment Sludge. Domestic wastewater treatment sludge analyzed pursuant to Fla. Admin. Code Rule 17-7.54(2) which meets the criteria for a Grade II sludge specified in Fla. Admin. Code Rule 17-7.54(3), and which has not been found to contain significant concentration of toxic substances,

may be applied to the land if the following conditions are met:

(a)   Application is limited to sod farms, pasturelands, forests, highway shoulders and medians, plant nursery use, land reclamation projects, and to soil used for growing human food chain crops (excluding root crops, leafy vegetables, vegetables to be eaten raw, and tobacco).

(b)   Use on playgrounds, parks, golf courses, lawns, hospital grounds, or other unrestricted public access areas where frequent human contact is likely to occur is restricted to processed domestic sludge and composted domestic sludge.

(c)   Total application amounts of Grade II domestic wastewater treatment sludge shall be restricted by limits set on heavy metal applications to agricultural lands. Maximum total heavy metal applications are (in pounds per acre):

> Cadmium - 4.45
> Nickel - 111.
> Copper - 111.
> Zinc - 222.
> Lead - 445.

(d)   Grade II - Application Rates.   No more than 10% of the total allowable amount of any criteria parameter shall be applied during any 12 month period.   In no case shall more than 500 pounds of total nitrogen or more than six dry tons of sludge solids per acre per year be applied to lands used for growing human food chain crops.

(e)   Grade II - Site Requirements
  1. Buffer Zone Requirements - The sludge land application area shall be located no closer than 3,000 feet from any Class I water body or 200 feet from any other natural or artificial body of water, except canals used for irrigation purposes or bodies of water completely within the site which do not discharge from the site.   The sludge land application area shall be located no closer than 300 feet from shallow private water supply wells, and occupied buildings (residences, offices, manufacturing facilities, etc.).   The sludge land application area shall be

located no closer than 500 feet from public water supply wells. Processed domestic sludge and composted domestic sludge are exempted from buffer zone requirements.

2. Soil Requirements - When application is to land used for production of human food-chain crops, the pH of the sludge-soil mixture shall be 6.5 or greater at the time of sludge application. When growing human food chain crops, testing of the pH parameter shall be done on a semi-annual basis, otherwise testing shall be done annually. Each site shall have a Conservation Plan which indicates that suitable soil infiltration rates and stormwater runoff control measures exist at the site.

3. Water Table Requirements - A minimum unsaturated depth of two feet is required above high water table level at the time of sludge application to the soil. Water table level shall be determined by observing the standing water level in a three-foot depth hole dug within the preceding 24 hours on the area to be used, or by measuring the water level in a monitoring well evacuated within 24 hours prior to measurement. No sludge land application may be conducted during periods in which surface soils of the sludge land application area are water saturated.

4. Special Requirements

a. Sludge land application area topographical grades must be eight percent or less.

b. The sludge land application area and an area 200 feet wide adjacent to and exterior of the sludge land application area boundary shall contain no subsurface fractures, solution cavities, or sink holes.

c. The general public shall be restricted from the sludge land application area for a period of 12 months after each application. Processed domestic sludge and composted domestic sludge land application areas are excepted.

d. Florida water quality criteria and standards as provided in Chapters 17-3 and 17-4, Florida Administrative Code, shall not be violated. Water quality testing of Grade II sludge application areas may be required if the department determines that sludge application not conforming to this section is evident. If water quality violations are indicated, the site owner shall cease any further sludge land application and have water quality tests performed as the department requires.

e. When applied to unvegetated soils, stabilized domestic sludge, other than processed domestic sludge or composted domestic sludge, shall be incorporated into the soil within 48 hours.

f. Depth of Permeable Soil - A layer of permeable soil of more than 2.0 feet thickness shall cover the surface of the land application area.

g. Pasture vegetation on which sludge has been spread shall not be cut or fed to livestock for a period of 30 days from the last application of sludge.

h. Sludge land application areas shall not be used for domestic animal grazing for a period of 30 days following the last application of sludge.

i. No human food chain crops except hay, silage or orchard crops shall be harvested from a sludge land application area for a period of 60 days following the last application of sludge.

j. No sludge shall be applied directly onto those portions of human food chain crops that are to be harvested and removed from the sludge land application area except hay on properly cropped and raked pastureland.

k. Domestic sludges shall not be used for the cultivation of root crops, leafy vegetables, tobacco, or vegetables to be eaten raw.

l. Stormwater runoff generated by storms up to a 10-year 1-hour event shall be prevented from entering or leaving the sludge land application area. Berms shall be placed for this purpose, if necessary.

m. Permanent records of actual application areas and application rates must be kept. These records must be maintained by both the site owner and the sludge land applicator for a period of five years, and must be available for inspection upon request by the FDER, Department of Health and Rehabilitative Services, or Local Environmental Program.

Sludge Records shall include:

(i) Date of sludge shipment or application.

(ii) Weather conditions when delivered.

(iii) Location of sludge destination.

(iv) Amount of sludge applied or delivered.

(v) Analysis of sludge pursuant to generator's permit.

(vi) Quality of sludge.

(vii) Sludge use agreements.

(viii) Area of land where sludge is going to be applied, location and designated future use of the land.

(ix) pH.

(x) Water table.

(xi) Topographic grades.

(xii) Vegetational status of sludge land application area.

(xiii) Incorporation of sludge.

(xiv) Depth of permeable soil.

(6) Disposal Criteria for Grade III Domestic Wastewater treatment Sludge.

All Grade III sludges and all Grade I and II sludges not to be applied to the land pursuant to Fla. Admin, Code Rule 17-7.54(4) or (5), and domestic wastewater treatment sludges not analyzed pursuant to Fla. Admin. Code Rule 17-7.54(2) shall be disposed of as follows:

(a) Sludge disposal shall be carried out at permitted or exempted solid waste disposal sites pursuant to Fla. Admin. Code Rule 17-7.03(2).

(b) Disposal of sludge shall be in compliance with Class I landfill criteria pursuant to Fla. Admin. Code Rule 17-7.05, except the covering requirements in Fla. Admin. Code Rule 17-7.05(1)(a).

(c) Sludge to be disposed of in conjunction with municipal solid waste (codisposal) shall be dewatered.

(d) Plans for sludge drying beds, sealed lagoons or other treatment facilities to be located at permitted solid waste disposal sites shall be submitted to the department as part of the permit application. Such drying beds shall not be located within 100 feet of municipal solid waste cells.

(e) Burning of sludge in volume reduction plants shall be pursuant to Fla. Admin. Code Rule 17-7.09.

(7)   Land Reclamation - Grade I and Grade II.

All Grade I and grade II domestic sludges may be used in land reclamation projects if the following conditions are met.

(a) Land Reclamation Application Quantity -Total application amounts are restricted by limits set on heavy metal applications to agricultural lands. Application of heavy metals may not exceed those figures listed in Fla. Admin. Code Rule 17-7.54(5)(c).

(b) Maximum total allowable application quantity shall be 30 dry tons/acre with such application to be accomplished within a one-year period on any given acre of sludge land reclamation area. The maximum allowable quantity shall not be altered by a change in the ownership of the land.

(c) The applied material shall be incorporated into the soil within 48 hours following application.

(d) Seed of a turf-forming grass shall be planted as soon as practicable, but in no case later than three months following cessation of sludge application.

(e) There shall be no occupied building (office, residence, etc.) or shallow drinking water supply well as defined in Fla. Admin. Code Rule 17-7.02(46) within 500 feet of the sludge land application area.

(f) There shall be no production of human food chain crops on the sludge land application area for a period of two years following cessation of sludge application.

(g) There shall be no surface water bodies within 500 feet of the sludge land application area.

(h) Public access shall be restricted for one year after sludge application.

(i) Topographical grades must be six percent or less before and after sludge application.

(j) Topographical grading shall be completed before sludge application begins.

(k) Stormwater runoff generated by storms up to a 10-year 1-hour event shall be prevented from entering or leaving the sludge land application area. Berms shall be placed for this purpose, if necessary.

(l) Florida water quality standards as established in Chapter 17-3 and required by permit in Chapter 17-4, Florida Administrative Code, shall not be violated. Water quality testing of Grade I sludge application areas may be required if the department determines that sludge application not conforming to this section is evident. If water quality violations are indicated, the site owner shall suspend any further sludge application and have water quality tests run as the department requires.

(m) Water Table Requirements - A minimum unsaturated depth of two feet is required above water table level at the time of sludge application to the soil. Water table level shall be determined by observing the standing water level in a three-foot depth hole dug within the preceding 24 hours on the area to be used, or by measuring the water level in a monitoring well evacuated within 24 hours prior to measurement. No sludge land application may be conducted during rain storms or during periods in which surface soils of the sludge land application area are water saturated.

(n) In addition to the above requirements, land reclamation projects at mining reclamation sites shall be in compliance with any applicable DER and Florida Department of Natural Resources rules concerning mining reclamation.
Specific Authority: 403.061, 403.704, F.S., Law Implemented: 403.702, 403.704, 403.705, 403.707, 403.708, F.S. History: New 6-16-84.

17-7.55   Criteria for Land Application or Disposal of Domestic Septage.

(1)   Discharge of domestic septage into permitted domestic wastewater treatment facilities shall be pursuant to the requirements of Chapter 17-6, Florida Administrative Code.

(2)   Application of domestic septage to the land, other than in

sanitary landfills as specified in Fla. Admin. Code Rule 17-7.54(6), shall be pursuant to Grade I requirements as described in Fla. Admin. Code Rule 17-7.54(4), if such sludge has been properly treated by lime stabilization, sludge pasteurization, or other process which produces similar kills of microorganisms.

(3)   Untreated domestic septage may not be applied to the land.

(4)   the disposal of domestic septage in sanitary landfills shall be pursuant to disposal requirements for Grade III sludges as specified in Fla. Admin. Code Rule 17-7.54(6).

Specific Authority: 403.061,   403.704,   F.S.   Law Implemented: 403.021, 403.061, 403.087, 403.702, 403.704, 403.705, 403.707, 403,708, F.S. History: New 6-16-84.

17-7.56   Criteria for Land Application or Disposal of Food Service Sludges.

(1)   Food service sludges may be discharged into permitted domestic wastewater treatment facilities pursuant to the requirements of Chapter 17-6, Florida Administrative Code.

(2)   Food service sludges may be applied to the land pursuant to Grade I requirements as described in Fla. Admin. Code Rule 17-7.54(4) if such sludge has been properly treated by lime stabilization, sludge pasteurization, or other process which produces similar kills of microorganisms.

(3)   Untreated food service sludge may not be applied to the land.

(4)   The disposal of food service sludge shall be pursuant to the disposal requirements for Grade III sludges as specified in Fla. Admin. Code Rule 17-7.54(b)(6).

Specific   Authority: 403.061,   403.704,   F.S.   Law Implemented: 403.021, 403.061, 403.087, 403.702, 403.704, 403.705, 403.707, 403.708, F.S. History: New 6-16-84.

17-7.57   Criteria for Land Application or Disposal of Processed Domestic Sludges.

(1)   Grade I and Grade II sludges which qualify as processed domestic sludges shall be land applied pursuant to the criteria of Fla. Admin. Code Rule 17-7.54(4), except that the records required by section 17-7.54(4)(d)4.1. can, as an alternative, be kept by generators instead of site owners or sludge land applicators and shall include only monthly summaries of (a) the amount of sludge delivered, (b) the amount of sludge applied, (c) the grade of sludge, and (d) a general description of the land where applied.

(2)   Generators of processed domestic sludge who produce such sludge in Florida or who deliver such sludge to Florida for use or final disposal in Florida, shall file a quarterly sludge analysis report and a quarterly sludge shipping and sales report with the Office of Solid Waste of the department.

(3)   Parameters to be tested for as part of the sludge analysis report are:

Total Nitrogen   - % dry weight
Total phosphorus - % dry weight
Total Potassium   - % dry weight
    Cadmium -mg/kg dry weight
    Copper   -mg/kg dry weight
    Lead   -mg/kg dry weight
    Nickel   -mg/kg dry weight
    Zinc   -mg/kg dry weight
Total Solids -%

(4)   Individual application rates shall be computed as follows:

Allowable lbs/acre/yr Cd - ppm Cd in sludge = lbs sludge/acre/year

- -

In the case of a 100 ppm Cd sludge and in a 20 year time span, this would be:

0.225 - 0.000100 = 2,250 lbs/acre/year of dry sludge.

A mixed commercial fertilizer using 10 % by weight of this sludge would have a maximum allowable application rate of 22,500 lbs per acre per year. Maximum application rates must be made available to sludge users by the manufacturer.

(5)   The agricultural use of processed sludge in accordance with this section is considered a normal farming operation and thus exempt from permitting requirements.
Specific Authority: 403.061,   403.704,   F.S.   Law Implemented: 403.021, 403.061, 403.087, 403.702, 403.704, 403.705, 403.707, 403.708, F.S. History: New 6-16-84.

17-7.58   Criteria for Land Application or Disposal of Composted Domestic Sludges.

(1)   Composted domestic sludge shall be disposed of pursuant to the requirements of Fla. Admin. Code Rule 17-7.54(6). However, such sludge which has been analyzed pursuant to Fla. Admin. Code Rule 17-7.54(2) and has been determined to be a Grade I or Grade II sludge pursuant to criteria specified in Fla. Admin. Code Rule 17-7.54(3) may be utilized according to the criteria provided in Fla. Admin. Code Rule 17-7.54(4) and (5), respectively.

(2)   Technical criteria for composting as described in Chapter 12 of EPA 625/1-79-011, "Process Design Manual for sludge Treatment and Disposal." This manual is adopted and made a part of this rule by reference. A copy of this document may be obtained by writing the department, and may be inspected at all DER offices.
Specific Authority: 403.061,   403.704,   F.S.   Law Implemented: 403.021, 403.061, 403.087, 403.702, 403.704, 403.705, 403.707, 403.708, F.S. History: New 6-16-84.

17-7.59   Effective Date.

This rule shall be effective 180 days after being filed with the Department of State.
Specific Authority: 120.54(12)(9), 403.061(7), F.S. Law Implemented: 120.54(12)(9), 403.021, 403.061, 403.087, 403.702, 403.704, 403.705, 403.707, 403.708, F.S. History: New 6-16-84.

Section 17-1.206 is amended to read:

17-1.206 Solid Waste.

    (1)   -   (3) No Change.

    (4) Land Application Field Package for Grade I Sludges.

    (5) General Permit Application for Grade II Sludges.

Specific Authority: 120.53(1), 403.061, F.S.
Law Implemented: 120.53(1), 120.55, 403.0875, F.S.
History: New 11-30-82.

Amended _____ .

17-4.64 is created to read:

17-4.64   General Permit for land Application of Grade Ii domestic
Wastewater Treatment sludge.

    (1)   As a result of the potential harm to human health or the
        environment resulting from sludge disposal activities, it is
        important that public notice be given before a general permit for
        such activities is utilized. Therefore, a general permit is hereby
        granted to any person for land application of Grade II domestic
        wastewater treatment sludge; provided:

        (a)   The person intending to apply the sludge to the land submits
            a completed General Permit Application for Grade II
            Sludges, as specified in Fla. Admin. Code rule 17-7.60.
        (b)   The permit applicant, within 14 days of notice to the
            department, has published in a newspaper of general
            circulation in the area affected, a notice of the intended
            land application of Grade II sludge. The notice shall include
            the name of the applicant and a brief description of the
            proposed activity and location.
        (c)   The sludge is land applied pursuant to the requirements of
            Fla. Admin. Code Rule 17-7.54(5).

    (2)   The general permit shall be subject to the general conditions of
        Fla. Admin. Code Rule 17-4.54.

        Specific Authority: 403.814, f.S.
        Law Implemented: 403.061, 403.087, 403.702, through 403.715,
        403.814, F.S.
        History: New.

Name of person originating proposed rule: Gregory L. Parker
Name of Supervisor or person(s) who approved the proposed Rule: Victoria J.
Tschinkel
Date proposed Rule approved: September 13, 1983

STATE OF FLORIDA

DEPARTMENT OF ENVIRONMENTAL REGULATION

Land Application Field Package for Grade I Sludges

This for is for utilization of Grade I sludge in normal farming operations pursuant to Chapter 17-7, Part IV, FAC. A completed form shall be submitted to the sludge generator by the sludge land applicator. A separate form must be completed for each sludge land application area.

No Grade I sludge land application project shall be initiated until the sludge generator has received a complete Land Application Field Package. Sludge land applicators and sludge generators shall maintain a permanent file of forms on sites actually used for sludge application. Such files must be made available for inspection upon request by the FDER or Local Environmental Program. Sludge land application areas must be made available for inspection by FDER, DHRS, or Local Environmental Program personnel upon request.

Use of Grade I sludge as specified in this form is exempt from solid waste permitting requirements provided in Chapter 17-7, FAC. Those persons who haul or use less than three cubic yards of Grade I sludge per month shall not be required to complete and submit the Land Application Field Package.

4.   Sludge Analysis (as required by Chapter 17-7, Part IV, FAC).

Analysis performed by: _____ Date _____
Address of laboratory: _____
_____

Test parameters for domestic sludge (weight figures in milligrams per kilogram dry weight).

(a)	total nitrogen	___%	(i)	pH	
(b)	total phosphorus	___%	(j)	solids	___%
(c)	total potassium	___%	(k)		___
(d)	cadmium	___	(l)		___
(e)	copper	___	(m)		___
(f)	lead	___	(n)		___
(g)	nickel	___	(o)		___
(h)	zinc	___	(p)		___
			(q)		___

5.   Residual Grade as per Chapter 17-7, Part IV: Grade ___

6.   Grade I application rate: gal/acre/year
                                   dry lbs/acre year

Calculation of maximum application rate:

8.34 (pounds per gallon) x (percent solids) x (percent nitrogen; dry weight) = pounds N per gallon

500 (maximum nitrogen loading; pounds) - pounds N per gallon

Sample calculation of maximum application rate (not to be used unless analysis indicates an identical situation)

8.34 x 0.02 x 0.045 = 0.0075
500 - 0.0075 = 66,666 gallons/acre/year
8.34 x 0.02 x 66,666 = 11,119 pounds (dry)/acre/year

No more than 500 pounds of total nitrogen shall be applied to one acre during any 12-month period. No sludges shall be applied to water saturated lands.

Landowners are responsible for water pollution or health hazards resulting from the application of sludge on their land.

Landowners must exercise sufficient control over the operation to insure that runoff, nuisance, or health problems do not occur.

All submitted data is accurate and complete. All required application site management practices shall be implemented.

_____   _____
Application Site Owner or   Date
Authorized Agent

_____   _____
Sludge Land Applicator   Date

Complete Land Application
Field Package Received:

_____   _____
Sludge Generator   Date

STATE OF FLORIDA

DEPARTMENT OF ENVIRONMENTAL REGULATION

General Permit for Grade II Sludges

This form is for utilization of Grade II sludge on agricultural lands and in land reclamation projects pursuant to Chapter 17-7, Part IV, FAC. Section I of this form must be completed by the sludge generator and the entire form must be provided to all sludge haulers, sludge land applicators, and site owners who use the generator's sludge.

The sludge land applicator of a Grade II sludge must complete Section 2 of this form and send a completed copy to the appropriate DER district office or approved Local Program. Land application of Grade II sludge shall not be initiated until notice has been given to the department at least 30 days before initiation of Sludge disposal.

Any proposed sludge land application area which does not meet the requirements for a general permit specified in FACR 17-7.54(5) shall require a standard solid waste disposal site permit from the department.

4.   Sludge Analysis (as required by Chapter 17-7, Part IV, FAC).

Analysis performed by: _____ Date _____

Address of laboratory: _____

_____

Test parameters for domestic wastewater treatment sludge (weight figures in milligrams per kilogram dry weight).

(a)	total nitrogen	___%	(i) pH	___
(b)	total phosphorus	___%	(j) solids	___%
(c)	total potassium	___%	(k)	___
(d)	cadmium	___	(l)	___
(e)	copper	___	(m)	___
(f)	lead	___	(n)	___
(g)	nickel	___	(o)	___
(h)	zinc	___	(p)	___
			(q)	___

5.   Sludge Grade as per Chapter 17-7, Part IV Grade _____

Section 3. Basic Site Date (Land Reclamation)

1. Name and current legal address of site owner or authorized agent:

    _____

    _____

    County _____ Contact Telephone No. _____

2. Land application area: _____ acres.

3. Sludge to be applied per acre: _____ tons (dry)

4. Productive capacity of land destroyed by _____
    _____

5. Name and address of County Health Unit licensed individuals or firms
    (if any) associated with the application project: _____

    _____

    _____ Telephone No. _____

6. Attach legal description of the site.

7. Attach letter of authorization for any authorized agent.

8. Attach a topographic map of the sludge land application area showing
    planned finished contours and any stormwater control facilities.

9. Attach a large-scale (small area) map of the site area. Indicate the
    following:

    A. Sludge land application area
    B. Nearest occupied building (office, residence, etc.)
    C. Nearest drinking water supply well
    D. Nearest state, federal, or county maintained road or highway.

INDEX

Acid formers 22
Aerated pile composting 143,174
Aerobic digestion 25,181
Aerotherm 144
Agricultural land, sludge disposal on 64,149
Air Lance 275
Air quality standards, Florida 106
Albert Whitted Wastewater Treatment Plant 107
Alkalinity 23
Ambient air quality standards 103,106
American Bioreactor 144
Americal Bio-Technology 144,277
Andco-Torrax 138
Anaerobic digestion 22,181
Ascaris 67

Bailey, S. 147
Bailie process 139
Barging sludge 154
Belt filter presses 40,55,202,301
B.E.S.T. process 136
Beta concept in centrifuge scaleup 50
Blinding in filters 55
Bridgeway Acres Landfill 100
Buchner funnel apparatus 58
Bulking agents in composting 34

Cadmium 65,83
Calorific value 30,33
Campbell, H. 8
Capillary suction time 60
Carbon to nitrogen ratio in composting 232

Carver-Greenfield process 136,177, 206
    costs 219,247
Centrifugation 40,48
Characterization of sludge 4,123, 150
Chemical composition of sludge 123,150
Chemical fixation 150,175
Chlorination 181
City/county incinerator 194
    See also Pinellas County
City owner/operator approach 192
Clean Air Act 80
Clean Water Act 79,101
Classification of sludge by FDER 104
Closed vessel composting 34
    See also In-vessel composting
Code of Federal Regulations (CFR) 101,102
Co-incineration 173,179
Combustion of sludge 137,173,179
Composting 34,142,174,181,183, 208,211,250
Concentration of solids in sludge 4,110,120
Contract disposal 72,188,209, 211,243
Contract hauling 188,209,211,243
    costs 281,302
Cost estimates 75,213
COST 68 16
Costing factors 211
Cyclonic furnace 140
Cylinder diameter in thickening 44
Cylinder height in thickening 44

Dade County FL 136
Decanter centrifuge 48
Deep well injection 150,175
Dewaterability 20
Dick, R. 8
Digestion
    aerobic 25
    anaerobic 22
Dioxins 30
Disc centrifuge 40
Disinfection of sludge 36,179
Disposal practices in the USA 147
Dissolved air flotation 40,45
Distribution and marketing
    206,207,208
Distribution of water in sludge 5
Drying of sludge 29,40,134,171,181,
    183,206,246
Dynamic olfactometer 18

Earthworm conversion 142,145,174
Electric furnace 140,173
Environmental Protection Agency
    70,83,84,101
Estimating project costs 216
Evaluation of alternatives 73,253
Explosions 30
Extraction procedure (EP) 83

Fairfield digester system 144
Federal construction grants
    regulations 82
Federal Regulations, Code of
    101,102
Federal Water Pollution Control Act
    79
Fertilizer
    from sludge (see Largo FL)
    value of sludge as 122
Filter leaf apparatus 58
Filter press 40,56
Filter yield 57
Filtration 40,54,202,301
Flash drying 30,40,134,172
Florida Administrative Code 103

Florida Department of
    Environmental Regulation
    (FDER) 102,309
    classification of sludge 104
    regulations on sludge disposal
        104,147,309
    Chapter 17-7 Regulations 309
Florida Power Corporation 100,281
Flotation thickening 40,45
Flow properties of sludge 7
Fluidized bed furnace (incinerator)
    30,140,173,177,183,208,209
    costs 226
Forest land disposal of sludge 149
Fuel requirements in incineration
    261

Gravity filter 40,41
Gravity thickening 39
Grease incineration 179,195,241,229

Hansen, J.A. 65
Hatfield, W. 8
Hauling 188,209,211,243
    costs 287
Hazardous waste regulations 82
Heating value of sludge 33
Heat treatment 40,181
Height in thickening 44
Henningson, Durham & Richardson
    Inc. 100,179,301
Higher heating value 33
Huekelekian, H. 5
Hydrocyclone 40

Implementation schedule 303
Incineration 30,98,181,209,241,281
In-vessel composting 142,174,178,
    183,208,211
    costs 232,243,250,277,285,306
Ionizing radiation for disinfection
    36,180,181
Iron Bridge Wastewater Treatment
    Plant 135

Jiffy Industries 226,248
Jones, G. 12

Kansas City KS 249
Karr, P. 5
Kavanaugh, M. 5

Lagoons 40,181
Land application 64,67,148,175,211
Landfarming 67
Landfill gas utilization 195,241,281
Landfilling 63,148,175
Land reclamation 149
Landspreading 64,67,148,175,211
Largo FL 30,99,135,246,301
Lime encapsulation
   29,146,174,178,183,209,210,251
   costs 237,243,283,302,306
Lime treatment 27,146,180,181

Manure 65
Marine Protection Research and
   Sanctuaries Act 80
Marketability of sludge 262
Mesophilic digestion 23
Methane formers 22
Metro-Waste 144
Modular starved air incineration
   139
Moisture content in composting 35
Molten salt process 139
Multiple effect drying
   136,172,177,183,206,247
   costs 219,277
Multiple hearth furnace (incinerator)
   30,136,174

National Environmental Policy Act
   81
National Pollution Discharge
   Elimination System 79,81
Nitrogen 64,232
Nutrients in sludge 64,122

Objectives of sludge disposal 186
Ocean disposal 70,150,175
Odor 15,18,25,29

Olfactometer 18
Operation and maintenance criteria
   256

Parasites 67
Particle size 5
Pasteurization 180,181
Pathogens 15,25,26,34,35,67,
   180,300
Paygro system 144
Payout period 275
Peirce, J. 147
Penetrometer 51
Perforated bowl centrifuge 40
Permitting requirements, Florida
   105
Personalized specific odor number
   18
Phosphorus 64,122
Physical properties of sludge 152
Picket fence rakes in thickeners 41
Pinellas County 98,103,179,209,
   279,281,301
Pinellas County Resource Recovery
   Facility 98,209,241,279,281,301
Pittsburgh PA 251
Polyelectrolytes 45,54
Portland OR 250
Pretreatment 128
Production of sludge 9,10,118,
   126,127
Purac system 144
PUROX system 138
Pyrolysis 141,173
Pyro-Sol process 139

Quantities of sludge 9,10,122
Quicklime for stabilization
   29,146,174,178,183,209,210,251

Radiation for disinfection 36,80,181
Rat-holing 41
Reacto-Therm process 139
Regulations
   federal 79,101
   Florida 104,147

Reliability 75,256
Resource Conservation and Recovery
   Act 80,101
Resource Conservation Company
   136
Rheological properties of sludge 8
Roedinger Lime Post Treatment
   237,284
Rotary drum filter 55
Rotary dryer 40,135,171,172,
   183,206
   costs 213
   Largo FL 246
Rotary kiln furnace 30,173

Safe Drinking Water Act 81
Sand drying beds 40,61
Selection criteria 256
Settleability coefficient 51
Sidestream treatment 261
Sigma concept in centrifuge scareup
   49
Single hearth cyclonic furnace 140
Slaked lime for stabilization 27
Sludge
   characteristics 20,112,123,128
   disposal alternatives 63,132
   disposal at St. Petersburg
      107,111
   disposal practice in the USA 147
   drying 134,172
   loading rate on land 300
   production 9,118,126,127
   stabilization 15,36,179
Sludge Volume Index 42
Sod farm 98,114,300
Solar powered drying
   136,172,177,183,207,248
   costs 224
Solid basket centrifuge 40
Solid bowl centrifuge 40,48
Solids concentration of sludges
   4,119
Solids flux in thickeners 42
Solids loading
   in aerobic digesters 28

in anaerobic digesters 24
in centrifuges 49
Solids retention time 22,24,28
Solvent extraction drying 135,172
Sorrento FL 136,248
Specific resistance to filtration 58
Spray drying 137,172
Stabilization of sludge 15,36,179
Starved air combustion 141,173
Stirring in sludge thickeners 41
Strip mines, disposal of sludge
   64,149
St. Petersburg
   98,103,107,110,124,186
Supernatant from anaerobic
   digesters 25

Taulman/Weiss composting system
   144,277
Technology assessment 170
Technology profiles 132
Thermophilic digestion 23
Thickening 40
Toroidal drying 135
Toxic Substances Control Act 81
Toxins in sludge 20,112,122
Toytown Landfill
   98,100,114,206,300
Transformation curves 73,253
Transportation of sludge
   151,153,176,188,209,211,243,287
Treasure Island Wastewater
   Treatment Plant 124
Tube thickener 40

Unconfined composting 143,174
User fees 275,307

Vacuum filter 55
Value engineering 281
Vermicomposting 145
Vermiculture 145
Vertical packed bed reactor 138
Viruses 67
Viscosity of sludge 8

Wastewater flows, projections 125

Water in sludge 5

Welfare of community criteria 258

Wet air oxidation 139

Windrow composting 143

Wright-Malta process 139